MODERN DAY NINJUTSU

OSCAR DIAZ-COBO

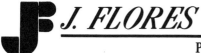

P.O. Box 14, Rosemead, CA 91770

© 1986 by Oscar Diaz-Cobo

Published by
> J. Flores Publications
> P. O. Box 14
> Rosemead, CA 91770-0014

All rights reserved. Reproduction in whole or in part without written permission from publisher is strictly prohibited.

ISBN 0-918751-04-7

Library of Congress Catalog Card No. 85-80885

Printed in the United States of America

Caution

This book has been published for entertainment and educational purposes only. Neither the publisher nor the author assume any responsibilities for the use or misuse of any information printed herein.

Preface

This book has been written as an introduction to the study of Modern Day Ninjutsu. It incorporates a blend of disciplines and military arts which are used to prepare for combat survival. Traditional fields of study reinforced with aspects of modern combat and survival maneuvers are presented.

Acknowledgements

The author would like to thank the following for assisting him with the illustrations in this book:

Bob Carbonell
Former U.S. Army (Airborne)
1st Dan Black Belt

Jeff Feagles
Former U.S. Marine Corps — Force Recon
1st Dan Black Belt
(Airborne & Ranger Qualified)

Eugene Floyd
7th Dan Black Belt

Steven Halasz
Former U.S. Army — Special Forces
(Airborne, Recondo, Dept. of Defense — Police Tactical Team
    Training
(S.W.A.T.)
3rd Dan Black Belt

Craig Mason
5th Dan Black Belt

Alfonso Ortiz
Former U.S. Army — Special Forces
(Airborne, Ranger, Jungle Expert)
4th Dan Black Belt

William Perez
Former U.S. Army — Infantry Scouts
1st Dan Black Belt

Raymond Van Boven
Former U.S. Army — Special Forces
(Airborne, Sniper, Scuba Qualified, Jungle Expert)
8th Degree Black Belt

The following Black Belts:
 Aldo Alvarez, Sihan Atlas, Carlos Diaz-Cobo, Carl Defronso, Durand Howard, Clint Lendor, Skip Meyers, Frank Miller, Roy Rohel

and   Clifton Martial Arts Supplies
        1157 Main Avenue
        Clifton, NJ 07011
For providing many of the traditional weapons which are illustrated throughout the book.

Table of Contents

Introduction	9
Physical Training	11
Heightened Sensory Perception	17
Hypnosis	20
Mental Force	26
Psychological Strategies	31
Invisibility	42
Misdirection of Attention	47
Chemistry	50
Combat Principles and Strategies	57
Unarmed Combat	74
Weaponry	93
Assassination	129
Sabotage	140
Espionage	142
Escape and Evasion	147
Conclusion	156
Appendix	157

Introduction

Ninjutsu is a specialized Art that can be employed as a means for individual defense in combat survival. Ninjutsu can also be employed by irregular forces for the purpose of special operations and/or to supplement regular forces. Traditionally Ninjutsu has specialized in the fields of espionage, assassination and sabotage. The covert skills of the practitioners of Ninjutsu were and can be employed to execute special missions such as an assassination to directly defeat the enemy. The practitioners were and can also be employed to indirectly defeat the enemy by gathering intelligence to offer the friendly forces an advantage in combat or to leak false information to confuse and mislead the enemy.

Government Intelligence Agencies and Military Armed

Forces have sections or forces that are involved with similar special operations. Similarities can be drawn between the traditional Ninjutsu roles and missions and the roles and missions of modern day Reconnaissance Marines, Army Special Forces, Navy Seals, and Intelligence officers and agents. The government forces and agencies have developed their special fields to a highly sophisticated and technologically advanced state.

This book presents the various training and disciplinary phases that should be engaged in to prepare for combat survival. The specialized fields of Ninjutsu are also presented. The preparatory subjects covered such as physical training, unarmed combat, weaponry, hypnosis, heightened sensory perception, invisibility, psychological strategies, mental force, escape and evasion and the specialized fields of assassination, sabotage and espionage that are presented in this book are covered to an effective yet rudimentary level. The individual practitioner should consider the material presented herein as a guide; and if he chooses to advance his degree of proficiency he should engage in in-depth professional training and study.

Physical Training

Physical fitness is very important to the modern day ninja. The ninja should strive to achieve and maintain a high degree of physical fitness. The missions that he may execute can be very physically demanding and stressful. By maintaining himself in excellent physical condition, the ninja will be better prepared to meet the demands of his missions. This section will present various means that can be used to achieve and maintain excellent physical fitness. The aim of physical training is to develop outstanding strength, stamina and endurance, to maintain excellent overall health and to prepare for specialized missions.

The following exercises, with their repetition standards or time limits, can be used as a guide for developing strength, stamina and endurance as well as for promoting excellent health.

Push-ups — This exercise is good for developing upper body strength. The ninja should strive to consecutively execute over 60 repetitions. Alternate the width of the positioning of the hands for varying effects.

Pull-ups — This exercise is good for developing upper body strength. The arms should be allowed to hang straight when in the down position and the chin should rise over the bar. The body may be allowed to gently rock as the repetitions are executed. Hand grips should be altered in width and facing forwards or towards the ninja for varying effects. Strive to consecutively execute over 20 repetitions.

Sit-ups — Sit-ups should be executed with the knees bent and the hands clamped behind the head. Strive to execute over 100 consecutive repetitions at a fast, steady pace.

3 mile run — Running is important for the ninja. Running improves overall health and develops endurance. Strive to run a three mile course in under 21 minutes.

The exercises of push-ups, pull-ups, sit-ups, and 3 mile run with the suggested repetitions and time limits can be considered basic physical training. The ninja should strive to excel in the basic exercises to achieve and maintain excellent strength, stamina and endurance.

The following exercises and training can be used to develop specialized power or they can be used to increase the ninja's level of fitness.

Run and Hide — The ninja can find this exercise useful. It can be used to develop excellent endurance and breath control. It is best accomplished in the woods. Run at a very fast pace for varying distances of 10, 25, 50 and 100 yards. At the end of each covered run, quickly slip behind cover. While under cover, the ninja should concentrate on controlling his breathing.

Run and Crawl — This exercise can develop excellent endurance. It can also be of use to the ninja. Run a distance of 40 yards then quickly dive to the ground and belly crawl 20 yards. Immediately following the belly crawl, rise and run a distance of 40 yards.

Both the Run and Hide and the Run and Crawl can be executed in sets or endured one after the other. Boots and suitable combat gear can be worn for motivation. The ninja can also make maximum use of cover as he runs by darting to and from large boulders or trees.

Run Circuits

The ninja can use run circuits to develop excellent endurance and stamina.

A run circuit can be as follows:

Run 50 yards and execute 30 push-ups, get up and run 50 yards and execute 50 sit-ups, rise and run 50 yards and execute 50 side straddle hops or jumping jacks, rise and run 50 yards and execute 50 bends and thrusts, then rise and sprint out 50 yards.

Run circuits can be executed in sets. Bends and thrusts are executed by squatting and assuming the up push-up position then rising.

Circuit Calisthenics

Circuit calisthenics can be used to develop excellent stamina. A calisthenics circuit can be as follows:

100 side straddle hops immediately followed by 100 bends and thrusts followed immediately by 100 alternate toe touches immediately followed by 50 push-ups followed immediately by 100 sit-ups.

The calisthenics should be executed at a fast pace with no rest between the exercises.

Calisthenics should be endured and executed to high repetitions. The training can develop excellent strength and stamina as well as promote flexibility and health.

Isometrics

Isometric exercises can be used to develop strength. The exercises should be executed with motivation and determination. The following are isometric exercises.

Palm to Palm — Press the palms together and exert maximum pressure for 15 to 20 seconds.

Clasp Hand Pull — Clasp your hands in front of you near the chest and pull outwards with maximum effort. Hold the pressure for 15 to 20 seconds.

Palm to Wrist — Press your hand to your wrist and exert

pressure for 15 to 20 seconds. Alternate between having your hand facing upwards and downwards. Also alternate top and lower hands.

Palm to Head — Alternate between placing and pressing your hands against the forehead, back and sides of the head. Maintain the pressure for 15 to 20 seconds. This exercise will develop the neck muscles to resist choking and blows.

Palm to Jaw — Alternate applying pressure to the bottom and sides of the jaw. Hold the pressure for 15 to 20 seconds. This exercise will develop and strengthen the jaw and neck muscles to withstand blows.

Weight Training

Weight lifting can be engaged to develop excellent strength. Weight training can be engaged with just a barbell, dumbbell and weights or with a wide variety of equipment and apparatuses. To develop strength, the ninja should exercise in proper form with consecutively heavier weights. By enduring intense workouts with heavy weights, the body can be strengthened and power can be increased. If a ninja chooses to lift weights, he must remember that his goal should be to develop and increase strength and power. He should not neglect other forms of physical training that will develop his endurance and stamina and that will help him to apply his strength in combat related events.

Dynamic Tension Breathing

This exercise is executed by bracing the feet at approximately shoulder width apart. As you inhale, relax the whole body and draw your hands to your sides. As you exhale, slowly tense your whole body concentrating especially on the abdominal region. Your hands or fists will slowly push forwards as you exhale. The exhalation should last between 5 and 10 seconds. You can thrust your hips forward to help tense the body. As you progress, try to bring the whole body into full tense in less time. Ideally, the body can be tensed with the air expelled immediately upon command. This skill will be useful when you receive punishing body blows.

Heavy Bag Training

The heavy bag can be used to develop strength, stamina, endurance, motivation and determination. It is an excellent training aid. When working with the bag, strike it with great force with the various striking techniques. Always try to hit the bag with more force and with greater penetration by using the proper pull back.

Swimming

Swimming is an excellent form of physical training. The ninja should attempt to increase his lung power and stamina by prolonged, steady swimming. Learn to swim strongly on and under the water.

Rope Climbing

Rope climbing can be a skill useful to the ninja. The proper technique should be learned. It involves using the legs as the major working muscles. Rope climbing can develop strength and stamina.

Combat-like Sports and Activities

Activities such as boxing, wrestling, Judo's randory, shadow fighting and Karate's full contact Kumite can be useful. These contact sports and activities can be engaged in to develop endurance, stamina and motivation. The contact sports can be useful to install a competitive spirit in the ninja as well as helping him become flexible and fluent in his motions.

It must be noted that although Judo, boxing, wrestling and contact Karate are excellent means for physical fitness and motivation, the sports cannot be considered a complete lethal form of hand-to-hand combat. Killing with the bare hands requires specialized training and discipline in the bare kill. Training in the sports can be used to supplement the ninja's skills.

Hand Conditioning

Hand conditioning should be practiced in two phases. The hands, fingers and wrist should be strengthened with gripping exercises such as squeezing a rolled piece of paper or rubber ball. The hands should be trained to develop a powerful squeeze and a strong resistance grip. The hands should always continue to be strengthened because powerful hands are a valuable asset. After the hands are strengthened, they can be conditioned for striking. A 2 by 4, padded with rubber, or a brick, padded with a towel, can be used as a makeshift striking post. A Karate Makiwara or a Gung-Fu Iron plam bag can be purchased for striking training. When training to develop powerful strikes, hit the pads only a few times with great force. Try to increase the force of your blows and not the repetitions. The edge hands, bottom fists, back fists, ridge hand, palm strikes, knuckle and finger strikes can be practiced and conditioned with the training pads. The training pads can be used to train the hands to tighten and position properly and to help overcome the fear of hitting. After the training session, the hands should be submerged and kneaded under hot water to increase the blood circulation and to decrease the chances of blood clots.

Specialized Combat Training

The following are specialized combat training maneuvers that can be used to prepare for combat survival.

Ping Pong to Forehead — A regular ping pong ball or small rubber ball can be bounced off the forehead to help develop eye-body discipline and coordination. Follow the ball with the eyes and attempt not to blink on contact.

Swinging Ball to Face — A rubber ball tied to a thin rope can be swung circularly very close to the face. Follow the motion of the ball while remaining steady and unblinking. In the advance stages of training, the ninja can catch or strike the moving ball while maintaining his composure.

The given physical training maneuvers can be used to forge the ninja into excellent physical fitness. The trainee should always remember to warm up before and warm down after a physical training session. The ninja must make demands on himself and endure tough, rugged physical training in order to condition his body for combat. In his motivated, determined training state, he must always be respectful of pain. Training should not be continued while in serious pain. The cause of pain must be treated before physical training is commenced.

An acceptable level of physical fitness should be achieved before continuing to the more advanced training.

Heightened Sensory Perception

Heightened sensory perception can be of great value to the ninja. By training and developing the senses to unconsciously detect specific stimuli, he can develop an unconscious security. The following are training exercises that can be used to heighten sensory perception.

Visual Sense

The ninja should develop his peripheral vision. He can do so by staring straight ahead and placing a training assistant to his left and one to his right. The assistants would then make subtle hand or foot movements which should be detected by the ninja without his eyes shifting. At a more advanced level, both assistants can move simultaneously and/or make subtle movements with their eyeballs to be detected.

The ninja should also develop his powers of scanning. He can scan a specific area paying very careful attention to detail. He can train by attempting to detect a small piece of glass or similar colored texture embedded in a carpet within a radius of two or three square yards. Scan the area from a distance of four to six feet.

Night vision should also be developed. He can do so by attempting to detect concealed objects such as weapons during periods of low visibility or at night. The ninja will find that vision at night is best when he scans using a ten to twenty degree angle or looking out of the corner of his eyes.

Audible Sense

The ninja should develop keen hearing. He can train by having an assistant breathe with varying force in a room. While just outside the closed room, he would attempt to detect the breathing and the location of the assistant.

The ninja should also develop selective hearing. This is accomplished by playing a radio at varying loudness and having an assistant silently approach him from the rear. He should attempt to detect the approach.

Tactile Sense

The ninja can develop his tactile sense by lightly touching or brushing an object. The objective is to attempt to feel the object exerting minimal pressure. Eventually, the object can be perceived without contact. The hand's static energy will alert the ninja of the object's presence.

Olfactory Sense

The ninja can develop his olfactory sense to detect smoke, perfume and body odor at varying distances. This skill will help him to detect campfires, fire or people.

Gustatory Sense

The ninja can develop his gustatory sense to detect unusual

tastes. In this manner he may be capable of detecting poisons or harmful additives.

The ninja should concentrate and focus his senses in training. By heightening his senses, he will be more apt to detecting vital stimuli.

Hypnosis

This chapter is designed to familiarize the reader with hypnosis and its potential uses to the modern day ninja. In order to understand hypnosis, the reader must be familiar with the dimensions of the mind. The dimensions of the mind are the conscious, the pre-conscious and the unconscious. The conscious mind is that in which the person is aware. The pre-conscious mind is that which is not in the conscious but can be recalled without the use of specialized techniques. For example, your name or your parents' names are usually not in your conscious mind but can be recalled with little effort. The unconscious mind pertains to the mental processes which can not be brought to awareness without the use of specialized means. The unconscious mind is responsible for activity or actions for which there are no apparent motives. The purpose of hypnosis or autohypnosis is to make an impression on the unconscious mind so that resulting activities or actions are favorable. For example, the ninja with the aid of hypnosis may make the suggestion that he will remain calm, alert and aggressive when engaged with enemy fire. If the suggestions are successful, the resulting action will be a calm, alert and aggressive state when engaged with enemy fire. The effectiveness of suggestions will depend on the training, conditioning, experience, confidence, the nature of the suggestion, his expectations, beliefs and reinforcing factors. Hypnosis holds no magical potentials. It is to be considered and used as a tool to aid the ninja in his self-development. Hypnosis is usually conducted with one operator and one subject. This chapter will concentrate on the uses of autohypnosis in which the operator and the subject are one and the same. Autohypnosis is a means by which to impress suggestions on the unconscious mind in order to achieve a favorable outcome. The favorable outcome can either be an almost immediate effect or a delayed effect.

Entry to Trance

A hypnotic trance can be defined as a very physically relaxed state with a very awake, alert and highly concentrated attention. In the beginning stages, it is best if the ninja makes use of the ritual entry to trance in a quiet, undisturbed room. Once he becomes adept, he can dismiss the ritual if he chooses and can enter the state even under unfavorable conditions.

The ninja should sit in a comfortable chair, his feet flat on the floor and his back straight. He may place his palms on his thighs or clasp them extending his index finger. The first phase will incorporate eye fixation and breath control. He should choose a focal point—a spot on the wall or the tips of his extended index fingers will suffice. Once the eyes are focused, the ninja should begin to breathe in a slow, rhythmic manner. With each breath, he should make a conscious effort to relax his body—allowing tension to flow away as he counts from 10 to 1. At the count of 1, the ninja should allow his eyes to close. He should then see in his mind's eye a light shade of grey. He will then slowly count from 10 to 1 and with each count, he will envision a darker shade of grey until at the count of 1, the color is nearly black. At this time, the ninja should be in a very relaxed and comfortable state. In effect, he is in a hypnotic trance, the depths of which will vary with individuals.

Do not make the mistake of believing that a hypnotic trance is a sleeplike state. It is not. The body may appear to be in a deep sleep, but the mind is fully awake and concentrated. If the ninja chooses to deepen his trance he can do so by concentrating on regulating his breathing or by counting backwards or by a combination of both. Deepening the trance may be helpful to gain confidence and focus concentration.

To exit the trance, the ninja should count from 1 to 10. When he reaches the count 10, he may open his eyes feeling fully awake, powerful and confident. Once proficient with this ritual, he may choose to come out of the trance instantaneously.

Suggestions

Verbal suggestions supplemented by the use of visualization are the principle tools or means by which to imprint the mind. Suggestions can be given to increase overall confidence and to intensify motivation and determination. Suggestions can also be directed at achieving a more specific goal. Suggestions can be given to cause a reaction to a specific situation. For example: the ninja may visualize a scene whereby he and the enemy are engaged in a shootout. Through visualization and suggestions, he can impress the attitude and reactions by which he will fire back confidently and skillfully to terminate the enemy. This type of positive suggestion helps to establish a frame of mind that will be suited towards handling dangerous, threatening situations. Suggestions do not replace experience. Suggestions, however, can be used to increase the level of proficiency and to heighten the effects of experience. Suggestions can be used to internalize other reactions, for example: to unconsciously check the hands of individuals that approach the caution distance, to react in an aggressive, skillful manner if the enemy's armed or threatening, to unconsciously note and record possible weapons and routes of escape. Suggestions can be used for even more specific results. For example: to hold the pistol in the proper grip and stance, or to execute a specific block to ward off a specific attack. Actions or reactions that are internalized and

consequently executed as a result of hypnosis training can be termed hypnotically induced reactions or reflexes. It will be noted that skill will be a result of practical training as well as hypnosis training. Hypnosis is used to create the scenery and to execute mental repetitions in order to intensify the effects of training and to properly set the attitude. The three phases by which to acquire the skill are as follows:
- learn to become aware physically or sensorily
- understand
- internalize

The skills can then be applied under varying and different situations through determination and mental flexibility.

Hypnosis can be used for creative thinking for the purpose of problem solving. In preparation, all information concerning the problem must be mentally recorded. After which, in a trance, the ninja would make the suggestion or command that his mind will work on the problem and it will arrive at a conclusion. The mind works in a progressive manner. Once the basic information is fed to it and a specific question or problem is asked, the mind will work utilizing its creative and logistic powers to arrive at a conclusion or answer. Although under certain circumstances, the answer may be arrived at almost immediately, it more often arrives in a post hypnotic state. The ninja would pose the problem and feed the known information. The mind would then arrive at a conclusion and the answer would filter down to the consciousness or awareness. Once the problem has been posed, it is best not to think about the situation again. The answer will come when it is ready, it is best not to force it.

Time Regression and Time Progression

Time regression and time progression are maneuvers that can be executed under hypnosis to return in time and review the past in detail or to progress in time and foresee events and the possibilities that can be encountered. Time regression can be useful to remember the past and draw from it valid information. Time progression can be useful in the planning of a mission. The technique of time progression can foresee possible mistakes or threatening events or activities that can endanger the mission.

The maneuvers are executed by entering a trance and either going backwards or forwards in time. The mind's eye would be used as a screen as the events are projected. The ninja has the power to freeze the projection to study it if desired.

Pain Management

Stress produces a series of chemical reactions and discharges in the body. Adrenalin is produced as a result of stress and is useful as an extra burst of energy. The adrenal glands pump adrenalin into the bloodstream to prepare the body to flee or fight. Adrenalin is responsible for what appears to be supernormal feats. As a result of stress, the chemicals endorphens are also discharged. Endorphens are a morphine-like chemical that serve as analgesics. They are natural pain-killers produced by the body.

The combination of adrenalin and endorphens is responsible for the continued fighting or running by an individual that has suffered serious, if not fatal, wounds. The adrenalin provides the energy-burst and the endorphens kill the pain.

In combat, an individual that has suffered serious injuries may continue his activity as a result of the chemical discharges and his total concentration and determination to accomplish his mission. The use of a battle-cry to increase the chemical discharges and to reinforce determination should be considered.

Once the individual has accomplished his mission and reached a place of safety and rest, he can follow the following steps to alleviate and manage pain.

1. The individual should assume a comfortable position.
2. He should then control, deepen and slow his breathing.
3. He should allow his body and mind to relax completely.
4. He can then concentrate on the specific area of pain and visualize the pain killing endorphens easing the pain as the natural healing process takes over.

NOTE: *First Aid can then be applied to protect the wound if necessary. First Aid maneuvers would be executed prior to the pain management procedures in the case of severe bleeding, choking, or burns.*

The individual's toughness and pain discipline will be an important factor in pain management. In the case of excruciating pain where the pain management procedures can not be exercised, demerol or morphine can be administered.

Mental Force

Mental force is the means by which to guide and control the natural energy or life force that is inherent in us. The procedures and information stated herein are based on psycho-physiological principles and natural laws. Knowledge is occult or mysterious only to those who do not understand. This chapter is designed to present the principles and uses of mental force. The study of mental force will expand the ninja's awareness and permit him to utilize more of his natural potentials.

Rituals
Rituals serve the purpose of preparing and motivating the ninja for his works. The ritual is conducive to a powerful, effective spell by creating a proper and concentrated frame of mind

Modern Day Ninjutsu 27

and a conducive atmosphere. The ninja has a broad range of choices that he can make to prepare or conduct his ritual. A special room, lights, medallion, candles, images, blood (his own) or the positioning of his hands for focus as in the traditional Kuji Kiri, can be utilized in the rituals. The ninja can make use of any of the artifacts for his personal preparations and works. He must, however, be aware that the artifacts are simple objects used to entice his own powers. He should never allow himself to believe that without his artifacts he can not perform his works. The power is from within.

Internal Control

Mental force can be used to control your internal functions. The keys to internal control are relaxation, concentration, knowledge, confidence, expectation and visualization. The following are techniques that can be used for internal control.

- The Switch

This technique is executed by relaxing the body and visualizing a switch much like a light switch. The switch is then flicked to the off position. As the switch is flicked to the off position, the ninja believes that the pain he was feeling has just been switched off. He visualizes the nerve transmissions being shut down in the area of pain or discomfort.

This technique is especially effective for the elimination of headaches and other minor pains. After the ninja has executed the maneuver, he should shift his attention to some other matter with a firm belief that the pain will soon eliminate itself.

- Blow Out

This technique is used to relieve oneself of tension or stress. It is executed by relaxing the body and then visualizing the tension gathering and collecting itself. The ninja then inhales deeply and exhales slowly and steadily. As he exhales, he visualizes the previously gathered stress or tension being blown out of his body and head.

The relaxation and visualization techniques function on the psycho-physiological principle that the body responds to mental thoughts and images. These techniques are of great value because by eliminating or easing minor discomforts and stress, the body is protected against more serious illnesses. High blood pressure, ulcers, cancer and a series of other illnesses can be avoided by practicing the given maneuvers. These illnesses are the results of a prolonged stress on the body and the consequent lowering of the body's defense mechanisms.

Somnambulism

Somnambulism is a very deep stage of hypnosis. In it the ninja can appear dead to the naked eye. The breathing is slowed to a non-perceptible degree and the heartbeat is so light and shallow that it is non-perceptible to the naked ear. The state can be used as a deep relaxation process or it can be used to create the illusion of death. Somnambulism is achieved by entering a state of hypnosis and then deepening the state with special emphasis on slowing the breathing process and heartbeat.

The Demon

This work consists of the development of a pre-set character designed for specific functions under varying conditions. It is a post-hypnotic work in the sense that it will manifest itself at required times without formal trance induction procedures. To create a demon, the ninja should enter a relatively deep trance. Prior to the trance induction, the ninja should make a mental list of the qualities that his so-called demon will possess. Ruggedness, determination, confidence, fearlessness and similar positive, productive qualities are encouraged. In the deep trance, the ninja will create an image of his demon. He will then attribute the chosen qualities to his creation. He will then visualize the demon expertly functioning in combat related scenario. Once the ninja is satisfied that the demon is perfect, he can visualize the demon entering his body, mind and soul. The ninja will then give himself the following suggestion or a very similar one: "Whenever I am in danger or I must accomplish a dangerous mission, the demon will become active within me and he will guide my thoughts and actions."

This work is centered on creating a separate personality which will manifest itself when required. The work is based on

Modern Day Ninjutsu

psychological and psycho-physiological principles. The demon and its functions can be compared to people who possess multiple personalities. The main difference is that the demon is consciously created and that the individual's personality is not submerged but rather enforced with the pre-set qualities that are advantageous and required for combat survival. The qualities are integrated at a heightened level into the personality through the use of the demon visualization and hypnotic suggestions.

External Influences

Mental force can be used to influence nature and others. Different works can be executed with varying degrees of success. External influences is based on the premise that man is a natural being in interaction and harmony with nature and subjected to its laws.

The means by which to accomplish external influences is thought projection. Thoughts have the natural compulsions to manifest themselves. In order to accomplish external influences the following factors must be adhered:

- Desire or Need

The first phase is to possess a great desire or need for the work. Without desire or need, the thought projection will lack energy or thrust.

- Clarity of Thought

When preparing for a thought projection, the thought must be clear, concentrated and focused. Random thoughts will not have the required composition and manifest. The ninja should have a clear image and result in mind to project.

- Preparation

To prepare for an external influence work, the ninja should be well rested and clear headed. Research on the specific target or area to be influenced can help the ninja use the proper visualization.

- Balance Factor

All works—internal and external—function through natural means. In order for a thought projection to manifest itself, there must exist a natural means for it to do so. This is referred to as the balance factor. It does not matter if all of the regulatory factors are satisfied—if the balance factor is not met, the thought will not manifest. The ninja has very little or no control at all over the balance factor. This is the cause for powerful thought projections, which have been expertly executed, not manifesting.

The ninja, being aware of the determining effects of the balance factor, will not rely on external influence works to accomplish his mission. The external influence works are executed for the following reasons:
- To increase the chances of the mission's success through natural influences
- To set a positive frame of mind with which to carry out the mission.

The visualization techniques that can be used to perform the external influence work are many and varied. The following are two different strategical approaches that can be used.

The Encompassing Cloud
This technique is executed by visualizing the thought energy encompassed in a cloud-like substance. The cloud is then thrust at the target where it releases the thought energy.

The Direct Visualization
This technique is executed by visualizing the mind and the target in a direct link. The thought is then focused and concentrated until it is completed.

In both techniques, once the thought projection is completed, the ninja should shift his attention to other matters. If the ninja continuously dwells on the thought projection, the force or thrust is diminished.

In closing, it should be noted that aside from the given advantages, the concept of mental force can be disguised and prostituted to bring fear to an uneducated enemy. It can be used to take advantage of the enemy's natural fear of the unknown.

Psychological Strategies

The ninja must endure tough, stressful psychological training to prepare for combat. Stressful missions and situations will demand a sturdy character and a command of the psychological strategies and ploys. The first section of this chapter will concentrate itself with the psychological and motivational combat training that should be endured by the ninja to strengthen and forge his character. The ninja should maintain a self awareness through introspection and constructive criticism. By remaining aware of his faults and strengths, he can best prepare to retain his means of confidence and enemy superiority. Knowledge of his faults can also motivate him towards correcting or improving his weaknesses, thereby improving his all around efficiency. Overall training must be tough and demanding and designed to ensure that confidence is deep rooted based on realistic abilities and practicality and not on superficial or misleading beliefs. The ninja should condition his mind to prepare for and to dominate dangerous situations. In the event that he is caught unprepared or overpowered, he should condition himself to continuously remain alert and prepared to take advantage of any event that can offer him the advantageous position or the chance of escape and evasion. The mind must be conditioned to disregard all de-motivating, subduing attempts and to remain in a confident, positive, active and alert state.

The following are psychological combat training maneuvers that can be used to train and mentally toughen recruits.

The Double Bind

A recruit is ordered by instructor A to immediately run to a designated room and to scrub the floor clean. The order is given that the floor must be scrubbed immediately and quickly. The recruit runs to the room and commences to vigorously scrub the floor. A few minutes pass when instructor B enters the room and is filled with anger because the floor is wet and soapy. Instructor B without permitting the recruit to explain his actions, commences to bellow out threats of death and dismemberment. Instructor B orders the terrified recruit to clean up his mess immediately or he will physically punish him. Instructor B must never allow the recruit to explain his actions. The now terrified recruit will most likely quickly scramble to clean and dry the floor. Within a few minutes instructor A walks into the room and begins to swear and roar in a rage of anger because his orders were not carried out. The recruit is never allowed to explain his actions and is belittled, threatened, confused and demoralized throughout the process. Eventually the recruit is ordered away with no explanation given. The recruit was caught in a double bind. No matter what course of action he followed he would be wrong one way or another. This type of training can be continued until the recruit calmly accepts the conflicting orders with no fear or confusion. When the double bind no longer causes aggravation or tension, the training has served its purpose.

The Unconscious Eye

This training exercise is designed to condition the recruit to keep his mind actively alert while suffering from fatigue and stress. The recruit should be physically exercised and deprived of sufficient sleep in preparation for this exercise.

The recruit is ordered to enter a room and to walk through it and out the rear door at a steady regular pace. The room can be made up to look like an office or a prison. Two instructors enter the room yelling and barking threats and abuses at the recruit. The recruit can be occasionally slapped on the back of his head or lightly shoved to help establish a stressful state. Throughout the walk the recruit is told to hurry up yet he is not permitted to increase his pace. The recruit is told that when he exits through the rear door he will be assaulted and beaten by awaiting instructors. The instructors must work towards terrorizing the recruit and focusing his concentration on the exit door. In the room various weapons or excellent makeshift weapons should be positioned in semi-concealed areas. Weapons such as a .45 calibre pistol can be placed on top of a desk semi-covered with a newspaper or a length of steel pipe can be placed to rest against a wall. The instructors must work towards creating stress, however, they must not position themselves in any way that will block the recruit's view of the weapons. When the recruit finally exits the room, he is ordered to attention and questioned on what he saw inside the room. He is then dismissed. The purpose of this exercise is to condition the recruit to disregard all abuses and threats of danger and to concentrate on spotting the weapons that could be used for survival. If the recruit spots the weapons and actively and quickly formulated a plan or various plans of escape and survival, then the training is completed.

This training can help condition the ninja to maintain an active survival oriented mind even under periods of great stress and fatigue.

The Wake Up

This training exercise is designed to condition the recruit to instantly wake from sleep capable of instant action. This training can be accomplished by having an instructor wake up the recruit at irregular periods and times and by forcing the recruit to physically exercise and by asking the recruit questions. For example: An instructor can wake up the recruit at 3 o'clock in the morning if the recruit went to sleep at 11 o'clock that night and quickly order him to grab his weapons and commence to run in place. As the recruit runs in place, the instructor can ask the recruit what his name is, what 7 x 8 equals, where he is and why, etc. . . . When the recruit can wake up with minimal stimuli and easily execute the physical training and answer the questions with a clear mind, the training is completed. The ninja can find it necessary to quickly wake from sleep and run for survival or engage the enemy in combat. This training will condition him to wake from sleep with an active mind and body.

The ninja can condition himself for an instant wake up with the aid of an alarm clock. He can set the alarm to ring two or three hours after he falls asleep and respond to the alarm in a rapid demanding manner.

The following are combat motivational training exercises. The training exercises are considered highly motivational because the trainee must conquer his fears and realization of danger and exhibit a disciplined professional attitude.

The Beheading Sword

In this training exercise a long sword such as the Japanese Katana or a machete is required. The trainee is told that the sword or machete will be swung at his neck in an attempt to behead him. He is allowed to assume a comfortable stance and position from which he can quickly and comfortably duck very low. A few very slow swings are executed so that the trainee becomes familiar and prepared for the potentially killing strike. The trainee is then issued a preparatory command and the sword or machete is swung at his neck with great speed and force. The trainee will have ducked well beneath the swing and will remain unharmed. The sword or machete's swing should create a fearsome shish-like sound as it was swung. The trainee must be thoroughly convinced that if he did not duck quickly he would at present be dead. The instructor must be certain that the trainee's duck will bring him to safety. The trainee is not allowed to move backwards or to any position except straight downwards. The swing is executed completely across and will not follow the trainee in any way. A loud battle cry can be used with the swing.

In Face of Death

For this training exercise a short sword such as the Japanese Wakizashi or a machete is needed. The trainee is told to stand at a position of attention. He is then told that the sword or machete will be swung very fast and very close to his face. He is to maintain his eyes open and not move or blink at all. The instructor then measures off the sword or machete so that it appears as though it will pass very close to the trainee's face. In reality, the sword or machete will be approximately six inches from the trainee's face. The blade will appear much closer to the trainee because of his position. The sword or machete is swung and the trainee is set at ease. A loud sharp scream can accompany the swing.

In both the beheading sword and the face of death training maneuvers, the instructor must be extremely capable with his weapon and have an intrinsic awareness about the trainee's level of tenseness and capability to respond.

Move or Die

For this training exercise, a few shurikens or large stars or knives capable of being thrown are needed. The trainee is told that he must quickly move to one side to avoid the weapons that will be thrown at him. His movements are first supervised so that he conserves movement and successfully clears his body and head out of the line of fire. He is to execute a quick side step and swing his body and head clear out of the line of fire. His starting position for each throw will be one that squarely faces the instructor. The instructor can first slowly toss a few stars or knives to familiarize and prepare the trainee. The trainee must then avoid the weapons that are quickly thrown at his mid-body. The last thrown is aimed and thrown at the center of the trainee's face. He is forewarned before the throw. The quickly thrown weapons should be limited to no more than 7 or 8. The instructor must be very capable with his weapons and be aware of the trainee's level of tenseness and capability for proper response. The trainee can be placed in front of a wooden wall so that the stars or knives stick to the wall leaving a stronger impression on the trainee.

Prisoner

For this training exercise, the subject and two assistants are needed. The trainee is told that the assistants, preferably trainees themselves, are going to securely hold his arms to his sides and rear. The trainee will then be beaten by the instructor. The trainee is not allowed to pull free or fight against the assistants or to attack the instructor. He is allowed to tighten up his body and to utter battle cries for strength and motivation. The trainee is then securely held in place and the instructor executes blows to the trainee's abdomen, muscular chest plates, thighs and slaps to the face. The instructor will execute no more than 15 to 20 semi to powerful blows. Immediately after enduring the punishment, the trainee is released and ordered to execute 25 consecutive push-ups. The trainee should concentrate on overcoming the pain by drawing on his reservoirs of strength and motivation. The trainee is then dismissed. The instructor must be cautious not to inflict injury on the vital targets of the trainee such as the ribcage, sternum or collarbone.

Circle of Death

For this training exercise four or more assistants will be needed. The trainee is surrounded by the assistants each maintaining approximately two yards distance from the trainee. Each assistant is secretly given a number. The trainee is told that when the number is called the assistant will execute a simple attack. For example: one lunge punch or one knife thrust. The trainee is to attempt to check, block, jam or avoid the attack and immediately counterattack. The trainee may not leave the circle. He may assume any fighting position he chooses. The trainee may not initiate any attack, he must wait to be attacked. The instructor will call out the numbers at random and supervise the training. The assistants can be periodically given changed numbers. In the advanced stages, weapons of wood can be introduced into the training. In such cases the blows and strikes must be executed with discipline and control. The trainees must be trained in the techniques of unarmed combat and in weaponry to execute this training phase. The trainee should attempt to get a feeling for the attacks and attempt to develop the combat sense.

In this training exercise, the instructor must fully dominate the class. He may have to discipline the trainees or assistants for lack of control and discipline.

The Will to Win
In this exercise two or three assistants are needed as well as full contact gloves. The trainee is told that he must combat the two or three fighters at the same time. All punches are allowed and elbow, headbutts, kicks, and knee strikes are permitted. The fight is commenced and should be kept in an aggressive state for a period of 2 minutes. If the fighters fall to the ground, they are ordered to rise as soon as the fight stagnates. The trainee is told to be as aggressive as possible and that his mission is to endure the oncoming punishment while inflicting as much as possible. The trainee must be in excellent physical fitness and possess outstanding endurance, motivation and determination. The instructor must fully dominate the class to keep it under control. The fight can be executed in a ring or a specified area.

The Psychological and Motivational Combat Training exercises can be used to develop the ninja's sense of personal toughness, self esteem, confidence and motivation. The training maneuvers are used to help build the well rounded, well trained, highly disciplined, highly motivated, tough fighting machines that are a breed apart from the common man.

The second section of this chapter is designed to familiarize the ninja with the psychological manipulatory tactics and maneuvers that can be used to overcome the enemy. The first phase towards psychologically manipulating the enemy is to do research or acquire knowledge about the enemy. The enemy is manipulated through his weaknesses. The enemy should be studied to know the intensity and range of manipulation that can be achieved through his emotions. The emotions being fear, hate, lust or love, envy and compassion.

The following characteristics or displays can be taken advantage of to manipulate the enemy:

A great desire for sensual pleasures. An enemy that wishes for a partner in sex or for sensual pleasures can be manipulated by employing an agent that will arouse and satisfy his cravings.

Temper bursts or quick irrational decisions and actions. An enemy that displays these characteristics can be manipulated by sparking the irrational bursts when detail planning is necessary.

Insecurity. An enemy that suffers from great insecurity can be manipulated by means of flattery and well placed compliments.

Compassionate or sympathetic weaknesses. An enemy that is overly sympathetic can be manipulated by playing on his compassionate tendencies.

Crave of adventure. An enemy that craves excitement can be manipulated by offering him the supposed chance of adventure and glory.

Laziness or lack of motivation. An enemy that is lazy and that has no desire to do some specific or non-specific job can be manipulated by offering the enemy a chance to relax while his job is supposedly carried out by another.

Fear or cowardness. An enemy that is a coward or that has a specific fear can be manipulated by the threat of danger.

Vices and habits. An enemy that has a vice such as the use of illegal drugs or alcoholism can be manipulated through his vice. An enemy's habit can be used to manipulate him because the ninja will know what to expect from the enemy when he is subjected to that stimuli. For example: If the enemy has the habit of patronizing the same lounge after work daily, then a meeting can be arranged so that the enemy is supposedly seduced by an agent.

The ninja must be proficient at lying and being deceitful to manipulate and take advantage of the enemy's weaknesses. Psychological manipulations can range from quick unexpected bluffs used at gunpoint to carefully planned delicate scenarios that take advantage of the enemy's need for ego gratification. The ninja should be aware of the manipulatory ploys and tactics so that he can take advantage of them as well as guard against them.

The ninja must make it a rule never to fear the unknown. The unknown is very often misinterpreted and can be a source or a means of manipulation. He can manipulate the enemy by playing on his unfounded fears or ignorance. The ninja will strive to discover the mechanisms or anything that can be used to manipulate—based on the enemy's ignorance or blind faith. The enemy's religion or superstitious beliefs can be a powerful means for manipulation.

The last section of this chapter will cover a topic which I call the steel mind. The steel mind is a combat trance that allows the ninja to remain totally unaffected by dangers or threats, and react in an extremely rapid, powerful, expertly and lethal manner. The steel mind thrusts the ninja beyond the bonds of human rationality or sanity. Factors that contribute towards the steel mind are knowledge, experience, confidence, technical and tactical proficiency, determination or oneness of mind, discipline and physical prowess. A person that has assumed the steel mind position is like the madman that transcends the bonds of rational power. The ninja that is in the steel mind position is also gifted with an extremely sharp calculating and reflexive mind that enables him processes which are beyond the reach of men in an average state. The steel mind is a psycho-physiological phenomenon in which a warrior may be suddenly placed into due to circumstances beyond his control. The ninja should strive to understand the psycho-physiological phenomenon so that he can make use of it at will when necessary.

The steel mind is unlike just an extra flow of adrenalin in that the psychological processes are also heightened and the behavior or actions of the warrior reflect to some degree his training and aspirations in training. The steel mind can be fairly described as a phenomenon which unleashes an outstandingly motivated, confident, determined, calculating to the point of reflex, expert warrior with an extremely high threshold to pain and total disregard for the negative effects of danger. Danger is, however, a powerful motivating factor for the steel mind.

Training for the steel mind encompasses the physical and mental trainings which are a part of the ninja's combat training.

The steel mind is a psychological and manipulative tool in that the warrior's spirit is projected and it serves to demoralize and even terrorize the enemy. The steel mind is the ultimate combat mental and physical state in which the individual is almost replaced by a pure state of highly coordinated, integrated physical, mental and spiritual expression of combat proficiency.

In closing, the ninja should train himself to instinctively utilize psychologically manipulating maneuvers and to take advantage of the enemy's confusion, surprise or weaknesses.

Verbal Abuses for Psychological Manipulation

This section presents some verbal assaults that can be used to intimidate, degrade, insult and provoke irrational behavior. The verbal assaults can be helpful in the training of a trainee to cause stress and to help toughen his character. The verbal abuses can also be useful to manipulate the enemy by insulting and frustrating him; causing him to become angry, therefore, provoking irrational behavior. The effectiveness of the verbal assaults depends on the following:

- The voice and character projection of the abuser
- The choice of abuses
- The circumstances and condition of the attempted manipulation
- The training and degree of professionalism of the intended victim
- The character of the intended victim
- The ability of the abuser to continuously verbally assault the victim and to elicit an emotional response.

The following are some verbal abuses that can be used to initiate the assault. The abuser should learn to execute his abuses in a convincing, continuous manner.

- You disgusting little turd
- You fat, worthless scum
- You skinny, slimy turd
- You shit-eating maggot
- You filthy, slimy, worthless fuck
- You queer, slimy faggot
- You're a piece of aborted shit, maggot

These insulting phrases can be used as examples of verbal abuses. Abusive assaults attack the victim's insecurities and self-esteem. The purpose of the phrases is to break down the victim or to elicit an emotional response.

The ninja must be aware of the proper uses of verbal assaults for manipulation. He should also know and realize when verbal abuses are not working and when verbal abuses are inappropriate. The ninja will not want to use verbal assaults to prepare and heat-up his enemy for combat unless there are underlying reasons involved.

Invisibility

This chapter is designed to present the concept of invisibility and to cover some of the maneuvers and considerations which must be kept in mind in order to operate undetected.

The basic concept of invisibility is to blend in and become a part of your surroundings. In whatever surroundings you're in, in order to operate undetected, you must look, move, smell, think and if necessary, speak as part of your surroundings with no trace of foreign or unnatural actions.

Camouflage

The art of camouflage consists of creating an appearance which blends into the background or surroundings.

- **Jungle camouflage**

To camouflage oneself for the jungle, the clothes, equipment or cases and accessories should be of a leaf-like pattern with drab olive green, brown and tan colors. The face, hands and other exposed areas should be painted in a similar fashion with irregular patterns.

- **Wood camouflage**

To camouflage oneself for the woods, the clothes, equipment or cases and accessories should be of a drab grey or black color. The face, hands and exposed areas should be colored in a similar fashion with irregular patterns.

- **Desert camouflage**

To camouflage oneself for the desert, the clothes, equipment or cases and accessories should be of a drab tan color. The face, hands and exposed areas can also be colored a drab tan. Although the skin color is similar to a drab tan, the flesh should be covered to prevent the sweat oils from reflecting.

- **Snow camouflage**

To camouflage oneself in a snowy terrain, the clothes, equipment or cases and accessories should be of a drab white color.

- **Urban camouflage**

To camouflage oneself in an urban setting, the color light grey is suggested.

- **Night camouflage**

To camouflage oneself for night operations, the color black or midnight blue is suggested.

Noise and Movement

The ninja should adjust his manner of moving according to the terrain, cover and visibility that is offered to him. Slow walking or crawling can be used to move when there is appropriate cover and concealment. Belly crawling can be used when there is little cover and concealment and during periods of low visibility. Crab-like walking can be used to move close to a wall remaining in its shadow or in a room with poor visibility to avoid tripping and falling. A quick sprint can be used to move from cover to cover when there is little or no cover and concealment. The sprint can also be used to close in with the enemy when there is little or no cover and concealment.

Silent movement is accomplished by carefully and slowly placing your feet and if applicable knees, hands and elbows on cleared patches of ground. If you are crawling, place your knees on the same clear patch where your hands were placed.

All loose objects should be taped down and secured. Wool is the most silent of texture. Cotton is next in line.

Cover and Concealment

The ninja should make use of trees, walls, boulders, bushes, cars, etc., to keep himself behind cover, and concealed. He must be disciplined so that he can remain silent and unmoving. Movement is the most readily perceived action. The ninja can conceal himself in trees, bushes, roof tops, sewers, or in concealed pits below the ground. Some positions offer cover and concealment, others only offer concealment. If the ninja is preparing for a fire fight, he should find both adequate cover and concealment.

If the ninja plans to infiltrate the enemy, he must not only look like the enemy, but he must walk, talk and use the same mannerisms as the enemy. He should also have an adequate cover, such as a realistic sounding name, position and purpose.

The final step in invisibility is to leave no traces or clues behind. A weapon, piece of equipment, signs of specialized weapons or skill can be used to trace the ninja. Motive and opportunity are also factors that should be disguised since they can be used as leads and/or support.

Modern Day Ninjutsu

A Word on Disguises

Modern Day Ninjutsu incorporates a series of disguises that can be used to aid the practitioner in remaining invisible. Disguises range from simple alterations to complex, advanced physical and psychological re-structures. The following are maneuvers that can be used to alter your appearance and personality.

Hair
The length, style and color of your hair can be altered to change your appearance. You can also grow or cut your mustache, sideburns or beard. Hair coloring dyes can also be used to change the colors. Hair pieces and wigs can also be considered.

Jewelry
Varying articles of jewelry can be worn to portray different characters. The ninja may choose to portray a married conservative by wearing a wedding band and a watch, or a radical liberal by wearing neck chains, wristbands and an earring. The jewelry worn should compliment the clothes.

Clothing
Varying articles of clothing can be used to portray different characters. The color, texture and style of clothes worn can alter appearances. Care should be taken that the shoes or boots worn match the clothes.

Contact Lenses
Plain or prescription contact lenses can be worn to alter the color of the eyes. The contact lenses can also be used to replace traditional eye glasses. Eye glasses can also be worn to alter an appearance.

Fake Tatoos
Professionally drawn imitation tatoos that can be washed off can be used to alter an appearance.

Plastic Face
Imitation scars or fittings that alter the cheeks and nose can be used to alter an appearance. In special cases, the face can be restructured through plastic surgery.

The following mannerisms can be used to portray different characters.

Limp

A light or severe limp can be used to portray an injury or physical defect. The use of a cane or crutches can add to the portrayal.

Right or Left Handed

The ninja can create the illusion that he is left handed if he is right handed and vice versa by holding his weapon or a pen ... in his chosen hand.

Voice

The voice can be changed by altering the pitch and speed. Fittings or cotton in the cheeks can be used to alter the voice. To further the illusion, the ninja can vary his choice of words and sentence structure. Stuttering can also be practiced to create the illusion.

General Mannerism

The ninja can alter his character appearance by changing his mannerism. He can choose to smile or remain serious, be talkative or quiet, forward or reserved, powerful and forceful or shy and meek.

Weight Control

The ninja can alter his appearances by practicing weight control. By drastically gaining or losing weight, the appearance can be altered.

Disguises can be used to portray different characters in missions where the illusion will be of value. Disguises are also useful in escape and evasion situations.

Misdirection of Attention

Misdirection of attention is an art which is appreciated and very much utilized in Ninjutsu. It is a valued strategy used to overcome the enemy's resistance. This chapter will present some of the principles and maneuvers that can be used to misdirect the enemy's attention.

Play on Fears
Human beings have natural and acquired fears. The ninja can manipulate the surroundings so that the enemy is forced to focus his attention on overcoming his insecurities and fears. He will then be vulnerable to attack. The following are fears that can be thrust upon the enemy.

Darkness

Man relies on his sight as his primary sense. He is usually uncomfortable in the darkness or in an area of low visibility. The ninja can make use of the darkness to misdirect the enemy's attention.

Loud Noise

A loud, unexpected noise can be used to momentarily startle and mentally unbalance the enemy. He is vulnerable to attack during that period of mental unbalance. A loud, startling noise can be the result of a shock grenade or a forceful stomp and scream.

Fire

Man is usually fearful of fire and will instinctively attempt to avoid it. The ninja can use fire to terrorize the enemy and manipulate him for an attack.

The Unknown

This heading incorporates a wide range of possibilities. Man is usually fearful of the unknown. An enemy that is up against someone or something which he does not know or understand will tend to be hesitant and fearful. The enemy that does not know what to expect is ill at ease. The ninja can create this belief or illusion to mentally unbalance the enemy and prepare him for an attack.

Prepare the Enemy for One Action Then Execute Another

The enemy can be manipulated so that his attention is focused on one thought and therefore made vulnerable to an unexpected attack. As an example: the enemy can be warned that there is a five-man ambush set for his patrol on the right side of the road ten miles from his position. He can then be ambushed one mile from his position from the left side of the road by twenty men.

Unexpected Shock Maneuvers

An unexpected, startling maneuver or action can be used to momentarily stun the enemy and make him vulnerable to attack.

- a loud scream
- turning a high intensity light on an enemy's face when in the dark
- giving the enemy a menacing demon-like stare
- speaking in a low, unemotional voice then switching to a loud, forceful tone
- throwing a concussion or flash grenade

These maneuvers can momentarily startle the enemy and offer an opportunity for attack.

The ninja should be aware of the possible actions and maneuvers that can be used to misdirect the enemy's attention and therefore, set him up for an attack. He should also train so that he will not fall victim to similar maneuvers.

Chemistry

This section presents the chemicals and formulas that are used in Ninjutsu to create seemingly magical effects. Flash grenades, smoke grenades, smoke screen and fire breathing are among the topics covered.

Some precautions which should be adhered to while handling pyrotechnic chemicals are as follows:
1. DO NOT SMOKE while handling the chemicals.
2. Always treat the chemicals as though they are extremely dangerous. They are!
3. If you're not certain of what you are doing—don't do it.

Flash Powder

Flash powders are used to create a sudden unexpected flash

of light. They can be used to temporarily blind the enemy and are particularly effective in the dark or semi-dark. The formulas are as follows:

White Flash Powder
Potassium Nitrate .... 1 part by weight
Powdered Magnesium .... 1 part by weight

Extremely Brilliant Flash Powder
Aluminum Powder .... 1 part by weight
Potassium Dichromate........... 1 part by weight

Flash and Smoke Powder
Gun Powder .... 1 part by weight
Powdered Magnesium .... 1 part by weight

The above three flash powders are the most common and useful. Should the ninja decide to make different color flashes the following formulas can be used:

Red Flash Powder
Powdered Magnesium .... 1 part by weight
Strontium Nitrate .... 1 part by weight

Green Flash Powder
Potassium Nitrate .... 1 part by weight
Powdered Boric Acid............ 1 part by weight
Powdered Magnesium .... 1 part by weight
Powdered Sulfur .... 1 part by weight

The chemical of powdered magnesium metal can also be thrown into an open fire to cause a white flash and flame.

Flash Paper
Flash paper is a treated paper that when lit will ignite and produce a fiery flash. Flash paper is used in conjunction with flash and smoke powders. It is essentially used as the powders ignitor. Flash paper can be bought commercially at magic supply stores. It is advised that, unless you are proficient with pyrotechnic chemicals and procedures that you buy the commercial flash paper. Making homemade flash paper is a dangerous process.

The formula and procedure for flash paper is as follows:

1. Prepare a solution of four parts concentrated sulfuric acid and five parts concentrated nitric acid.
2. When mixing the acids do so slowly and in small quantities so as to avoid splashing or the build-up of heat.
3. Rubber gloves should be worn.
4. Place the solution in a shallow glass dish and soak white tissue paper in the solution for ten minutes.
5. Remove the paper from the solution using a glass rod.
6. Wash the paper in running water until all of the acid is removed.
7. After the acid is removed, dry the paper in the open air away from any sources of fire.
8. The paper can then be stored between cardboard or untreated tissue paper.

Flash Paper Pistol

A flash paper pistol has been manufactured. The device fits in the palm of your hand and is attached to your fingers. The tube is loaded with flash paper. By pressing a button with your thumb a small incendiary device is activated within the tube and causes the flash paper to ignite and to propel forward. At close range the flash paper pistol can be used to blind and sear the enemy's face.

Smoke Screen

A ninja may find it necessary at times to throw a thick smoke screen to assist him in remaining invisible. Smoke screens can be used in escape and evasion, infiltration, as a ruse, and also as a signal.

The chemical TITANIUM CHLORIDE when exposed to air creates thick volumes of smoke. It can be used to create a smoke screen.

Liquid Fire

Liquid fire is a solution which will ignite when exposed to air. It is very dangerous, highly flammable and must be handled with extreme caution. The formula consists of yellow phosphorous dissolved in carbon disulfide. It should be made fresh since the solution is not stable and it should be stored in a glass stoppered bottle because the carbon disulfide can dissolve rubber.

When exposed to air, the carbon disulfide evaporates rapidly and the remaining phosphorous ignites.

Liquid fire can be used to set aflame an enemy at close quarters.

Fire Breathing

Fire breathing or spurting flames from the mouth is a dangerous technique that should be executed only by those who are thoroughly prepared. Some precautions which should be observed are as follows:

1. Coat the mouth thoroughly with saliva.
2. Always exhale the breath continuously through the mouth.
3. If the lighted material burns the mouth, shut the mouth tightly and breathe through the nose. This will extinguish the flame.

The techniques of spurting flames from the mouth are as follows:

1. Lighter and lighter fluid technique

    This technique requires that you place lighter fluid in your mouth, light a cigarette lighter or strong match and maintain it about a foot or so in front of and just below the mouth level as you blow out the lighter fluid. The lighter fluid will catch on fire and blow out as a flame.

2. Brimstone technique

    This technique requires that you wet a piece of brimstone with gasoline, light it and place it in the mouth. When you exhale (through the mouth only) a vigorous flame will come out.

    Flames from the mouth maneuvers can be used in close to blind and sear the enemy's face.

Fire Proofing

Fire proofing is a skill which the ninja may find useful. To fire proof cloth material, dissolve a teaspoon of boric acid and

Fire breathing is one of the many specialized skills used by the Ninja to defeat his enemies.

a teaspoon of borax in 4 ounces of water. Saturate the cloth and allow to dry. To fire proof clothing larger volumes can be made.

Flash Bombs

The ninja can construct a flash bomb by enveloping flash powder in flash paper. The paper can be ignited and thrown to produce a flash in close quarters.

Smoke Bombs

The ninja can construct smoke bombs in a number of ways. The most effective and practical is to use a military smoke grenade. The ninja can also encase the chemical TITANIUM CHLORIDE in a plastic egg-shaped container. When thrown, the container would break producing a smoke screen. This is in effect an impact smoke bomb.

A flash smoke bomb can be constructed by enveloping the flash-smoke powder in flash paper. The makeshift grenade would then be lit and thrown to produce a flash and smoke at close quarters.

Methods of Igniting the Flash and Smoke Bombs

Flash and smoke bombs that are constructed and encased in flash paper can be ignited as follows:

1. Fuse — a fuse can be inserted into the mixture leaving one end protruding. In this manner the flash and smoke effects can be delayed by the fuse.
2. Direct fire — the flash paper can be ignited with a lighter, a match, or a cigarette. This method will produce an instant reaction.
3. Advance ignition systems — advance ignitions rely on chemical interactions which cause fire. The following chemical interactions will cause a fire:
   1. Liquid fire — previously covered.
   2. Sodium peroxide (1 part) mixed with granulated sugar (1 part) when mixed with water will cause a flame.
   3. Potassium Chlorate (2 parts) mixed with granulated

sugar (1 part) when mixed with sulfuric acid will cause a flame.

Using these advanced ignition systems an impact grenade can be constructed to cause fire (using a gasoline and oil mixture), smoke (using smoke powder), and/or a flash (using flash powder).

Ninjas skilled in chemistry, pyrotechnics, encasing techniques and chemical release units can construct potent impact or delayed action fire, smoke or flash grenades.

If you are not skilled in the above arts, you should not attempt to make smoke, flash or fire grenades using the advanced ignition systems because the chemicals are very volatile and dangerous.

A very advanced smoke grenade which can be constructed by experts in laboratories is the CACODYAL IMPACT GRENADE. Cacodyal is made by chemically extracting all of the oxygen from alcohol and replacing it with metal arsenic. Cacodyal will spontaneously enflame when exposed to air and as such it is very dangerous. The solution cacodyal when used in an impact grenade will explode and release a dense white smoke screen. This smoke is in reality white arsenic gas which is a deadly poison. If a ninja were to use this impact grenade, it would not only conceal his escape but it will probably disable or kill his pursuers.

Combat Principles and Strategies

This chapter is designed to familiarize the reader with the basic combat principles and the more advanced combat strategies. The combat principles form the base upon which Modern Day Ninjutsu maneuvers are founded. The combat strategies presented are means by which the ninja can control and manipulate his opponent in close combat.

Combat Principles
The principles presented herein are considered guides for personal defense and combat:

Preparation in training
The ninja must prepare himself for the hardships of combat

through realistic, arduous training. He should continue training throughout his lifetime continuously attempting to perfect his skills and acquiring new knowledge that can be of value.

Flexibility
The ninja should not train himself to function in only one manner or for one purpose only. He should train with flexibility in mind.

Adaptability
The ninja should train himself so that he can adapt to different conditions, circumstances and situations.

Confidence
Ninjas should develop and maintain a high degree of confidence through realistic training, tests and experiences.

Determination
The ninja should retain a focused state of mind irregardless of the difficulties he may encounter.

Ruthlessness
The ninja should cultivate a sense of ruthlessness for combat. Any hesitation or half-hearted maneuver can be the cause of his death.

Deceptiveness
The ninja should condition himself to act in a deceptive, unpredictable manner for combat. The enemy should be kept uncertain as to the ninja's next action.

Overkill
There is no such term as overkill in Modern Day Ninjutsu. Overkill is replaced by the term insurance. In combat, there exists no such actions as overkill. Seemingly overkill maneuvers are used to insure the success of the mission.

Terrorize
The ninja should attempt to break the enemy's fighting spirit whenever possible. Terror tactics can be used to bring fear into the enemy and to cause disheartening effects.

Victory without violence
This is the highest principle for combat. The goal of combat is to defeat the enemy. If the enemy can be defeated without

Modern Day Ninjutsu

a violent encounter, then the ninja has exercised the highest of all combat principles. He should attempt to dominate the enemy either through coercion or preferably through deception. In this way, the enemy may be used to serve his purpose.

Combat Strategies

The following combat strategies are maneuvers that can be executed to manipulate and control the enemy.

Blending Strategy

The blending strategy is an approach that could be used for both offensive and defensive purposes. It relies on body movements which allow you to flow through your opponent avoiding or re-directing his resistance. As you blend in with your opponent you can launch attacks. Your opponent is set up so that he can not effectively block your attacks. His only defense is to flow with your movements in an attempt to out maneuver you or to disengage. The following are examples of blend in maneuvers. They can be used in the offense or defense-counter-offense with the proper modifications.

Circular Blend

This maneuver requires that you redirect your opponent's initial attack or that you get close to your opponent and engage his guard as the first step. Once you are in close to the enemy you spin around him. The spin can be executed either towards the opponent's front or rear. The spin to the rear is safer since it lessens the chance of encountering the enemy's guard. If you spin to the enemy's front you can use 1) an elbow attack to the face, or midsection; 2) a forearm edge hand attack to his face or throat or; 3) by keeping low you can execute a bottom fist to the enemy's groin. Only practice and conditioned intuition will allow you to know the approximate distance of your enemy and whether a forearm edge hand strike or an elbow strike is appropriate. You must keep in mind that the enemy's guard may be encountered and therefore you must continue your spin after launching the initial attack in order to continue your attack.

If you spin to the rear you will roll across the enemy's back. You can 1) execute a forearm-elbow smash to the back of his neck or 2) execute an elbow smash to his spine or kidney region. You must then continue your spin and your attack.

Linear Blend

The linear blend does not require that you spin. In this maneuver you are required to get close to the enemy to engage and redirect his lead guard or to redirect his initial offense. You must then adjust your body to angle the enemy's. You can angle towards his front or rear. If you angle towards his front you can execute a front kick to his groin or a hand strike to his upper gate targets. If you drop low when you angle you can thrust a punch at his groin or execute a groin grab. If you blend in towards his rear you can execute a hook punch to his jaw, temple, or kidney, or execute a forearm smash to his base of the skull. A continued attack can follow.

Invisibility Blend

If you wish to extend the linear blend into the invisibility blend you can do so by continuing your motion past the enemy's rear. From this position you may crouch and execute a groin grab on the enemy. Your free hand would jam his arm and your face would be thrust against the enemy's back and pointed towards the safe side—the side where the enemy's arm is trapped. An alternate technique is to reach forward and grab the enemy's eyes with both hands and pull him backwards. You can gouge the eyes by maintaining his head against your center or you can bring the enemy down and breck his neck using the neck crane.

If you choose to pass the enemy altogether you can then latch on to his shoulder and reversing your momentum execute a punch to his jaw or temple. You could also execute a ridge hand or ridged forearm smash to his throat. A continued attack can follow. The invisibility blend is called so because you remain outside of the enemy's vision while executing your attacks. As such his best effort of attack will be blind techniques. The enemy may also attempt to outmaneuver you or disengage.

The blending maneuvers, once mastered, will offer you a viable option to offense and defense. It should be noted that

Modern Day Ninjutsu

the blending maneuvers can be executed efficiently unarmed or, with modification, with weapons. The procedures remain constant, you simply have to adjust the distances or attack techniques to meet the weapon's demands. The maneuvers should be mastered unarmed at first and then trained in with the close combat weapons of choice.

A precaution with the blending techniques, specifically the circular blending maneuvers: you must remain close to your opponent in order to execute the blending maneuvers. It is argued that the circular blending maneuvers are best executed as defensive-counter offensive maneuvers rather than offensive maneuvers. This is so because the trained enemy will have a conditioned response of pulling back and adjusting his stance and guard. If you are not in-close to the enemy, your circular blending techniques will become simple spinning attacks, which can be blocked by a trained opponent. When executing the blending circular maneuvers from the offense you must be certain that the enemy's guard is neutralized and that the enemy is stable so that he can not adjust himself prior to your execution of the techniques.

The blending maneuvers can be useful when you wish to avoid a prolonged fight and your primary purpose is to escape. If you are running and are faced by an enemy guard, you can run towards him, engage his guard and execute a blending maneuver causing injury with the initial strike. You can then continue your escape and evasion actions.

A particular technique that can be used when encountering an enemy barring your escape is to run towards him, push aside his guard as you execute a linear blend. In a continuous motion without losing momentum you can place your closest palm on the enemy's chin and with your other hand grab his hair. Then as you continue your run snap his neck forcefully back and to the outside. This technique can break the enemy's neck. The technique can be modified. Instead of placing your palm on his chin you can thrust your fingers into his eyes and then rake his eyes as you continue your run, forcing him to the ground.

Disabling the Guard

Disabling the guard is a tactic which is adaptable to unarmed

1

2

3

Circular Blending techniques. 1) On-Guard position 2) Parry strike and turn, 3) roll across enemy's back and deliver elbow to spine. 4) Execute jumping take-down and 5) deliver hammer fist to groin.

Linear Blending techniques. 1) Redirect your opponent's initial offense 2) Drop low and execute a groin grab 3) Blend toward his rear and execute a groin crush.

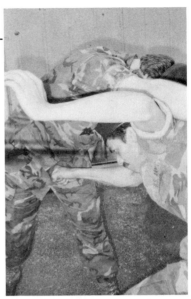

Low Linear Blending. 1) Crouch and block blow 2) Deliver punch to groin 3) Execute wrist and elbow lock 4) While simultaneously kicking throat.

Modified Linear Blending. 1) Assume On-Guard position 2) Engage and disable enemy's guard 3) Turn in and grab groin 4) Execute a sweep to bring your opponent down 5) Deliver a hammer fist to groin.

4

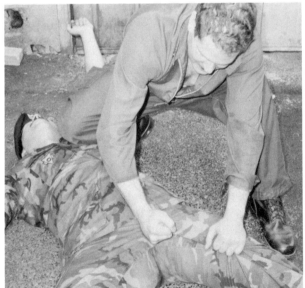

5

combat and to close combat with close quarter weapons. The main thrust of this tactic is to neutralize the enemy's guard; thereby neutralizing his offensive, defensive and counter-offensive capabilities, and to open a gateway for a powerful attack of your own.

There are numerous and various ways to disable the guard. For the purposes of this book, we will concentrate on the two primary methods: Trapping the guard and battering the guard. The more advanced maneuvers of occupying, distracting, misleading or entangling the guard are beyond the scope of this book.

Disabling the guard is a preparatory phase to the attack. A skilled fighter does not attack without first having set-up his opponent. If you simply attack without preparation you run the risk of having your attack upset by the skilled or instinctive reactions of your opponent, and unless you are skilled in the more advanced maneuvers of guard disabling you are back to step one, or worse, you become the recipient of the enemy's counter-attack. If you are in combat with a skilled adversary you should be aware that the enemy could be "drawing your attack" in order to set you up for an effective counter-attack.

It is true that action is faster than reaction. However, the action required to effect a block, avoidance, or parry is much less than the action required to effect a full attack. This is so because of the economy of motion which exists in effective defensive maneuvers. The tactics and maneuvers presented herein are designed to offset the defender's economy of motion advantage by restricting his motions or by neutralizing them altogether.

Concepts which tie in with the guard disabling tactics are the initial strike and the consecutive attacks. You must know what target areas you are making vulnerable, what attack you will execute and what attack or maneuvers you can utilize in the event that the enemy executes a partial defense despite your attempt to neutralize his guard.

A guard consists of the arms and hands and also the body positioning; the legs, shoulders and head positioning. In effect it is the whole body's position which constitutes a guard. We will concentrate primarily on neutralizing the enemy's hands and arms portion of the guard. If you neutralize the enemy's hand and arm guard, and attack skillfully and in a consecutive manner, you will overcome the other portions of the guard. It should be noted that if the enemy is skilled in the art of blending maneuvers you can expect difficulties.

I have covered attacks and maneuvers which can be used in conjunction with guard disabling maneuvers. It should be noted that guard disabling maneuvers can and should be used in conjunction with blending maneuvers in order to execute effective "complete attacks."

Trapping the Guard

This consists primarily of engaging and securing, i.e., trapping and holding, the enemy's forward arm. It should be done with your forward hand. By trapping the enemy's forward guard you have neutralized his primary defense and have effected some control over him.

A powerful hand strike can follow with your free hand to his exposed upper gate targets or to any other target region which the enemy may expose through movement. It should be noted that you are not limited to hand strikes.

Different variations of guard trapping exist. For example: you can initially trap the enemy's forward forearm with your forward hand and then switch it to your rear hand while you execute a strike with your forward hand. There are some circumstances in which you would bypass his forward arm and trap his rear arm in order to execute your attack. These variations exist but in general it is usually your forward hand to his forward arm.

When using close combat weapons the same principle applies. You trap the enemy's forward arm and you attack his exposed targets with your rear armed hand.

Battering the Guard

Battering the guard consists of striking the enemy's forward guard arm in such a manner so as to bring about great pain and temporary paralysis of that limb. As a result the enemy will not be able to utilize the injured limb for defense. A skilled quick and powerful attack must immediately follow in order to take advantage of the temporary guard disablement.

The primary weapon when unarmed for battering the guard is the fist. Skilled attackers, with conditioned hands, can use knuckle strikes.

There are numerous points on the guard arm that are vulnerable to attack. The ulnar bone, brachial bone, wrist, back of the hand, fingers and elbow are all bony structures which can be attacked and fractured to cause paralysis. These targets, however, are best attacked with a close combat weapon such as an entrenching tool or a pipe. They are not ideally suited for attack by the bare hands while engaged in battering the guard.

There is one section which is ideally suited for attack by the fist or by knuckle strikes. It is located approximately one inch out from the elbow and approximately three inches up towards the tricep. Sharp punches or knuckle strikes to that region will cause damage to the triceps' common tendon and the musculospiral nerve and will result in temporary paralysis of the limb. This region is ideally suited for attack when battering the guard because it is exposed in most guard positions.

When attacking this region be certain that your fist is turned in such a manner so as to allow for maximum penetration. Usually a fist with the thumb side down is superior.

Drawing the Attack

Drawing the attack is a strategy that is adaptable to both barehanded and close combat using close quarter weapons. The strategy is based on the principle that all guards leave openings and it is generally those openings which are attacked by the enemy. Using this strategy, the enemy's attack is controlled even before it is launched.

The ninja must be aware and proficient with the advanced strategies of guard disablement and blending. Guard disablement and blending assaults are two strategies which can be encountered when drawing the attack and one must be prepared to overcome their effects.

To draw the attack one must be aware of what target regions are vulnerable with respects to varying guards. The ninja must be aware of what targets he is leaving open to attack in a stable guard position and he must be especially sensitive to what areas he leaves open to attack while in motion. A knowledge of defenses and counter offenses must exist. More important than the particular defenses or counter offenses which can be used are the strategies or mind frames which motivate the maneuvers. The ninja must perceive the attitude which the enemy possesses when stable or in motion. Reading your opponent and knowing and understanding his attitudes is a skill which is of inestimable value to the ninja. Although the more advanced reading or "radar" techniques are beyond the scope of this book, the basic attitudes which you are looking for are as follows:

Will the enemy, when approached, withdraw, advance or stand his ground; and will he attack by probing or will he lunge into a committed attack.

Depending on the enemy's attitude and actions you will have to adjust your ploys of drawing the attack and for executing effective counter attacks.

There are two principle methods for drawing the attack. They are *inviting the attack* and *forcing the attack*.

Inviting the Attack

This is executed by engaging the enemy or by having him engage you while you maintain a guard which has a penetrable flaw. This is not to say that the guard is useless, only that it will leave an opening which will be perceived by the enemy and to which there is a reasonable belief that he will elect to attack through it. Varying angles and hand and arm positions will effect different openings. You must be aware of the openings and be prepared to reflexively neutralize the enemy's attack and quickly counter. It must be noted that the major opening that you leave will not preclude the enemy from electing to attack you through alternate means. You must remain sensitive to the enemy.

Forcing the Attack

This is executed by assaulting the enemy in such a manner so as to force a defensive maneuver and to elicit a counter attack to an exposed target region. You must be prepared in turn to reflexively neutralize his counter and to deliver a counter attack of your own.

These concepts may at first seem difficult and complex, but through training you will find that they can be easily incorpo-

Drawing the attack. 1) Author assumes a guard which exposes his knife hand. 2) His opponent attempts a hand cut; Author retracts his knife hand 3) and counters with a trap and knife thrust to the throat.

rated into your arsenal of fighting maneuvers. It will be noted that master fighters execute these strategies instinctively and with no cognitive process involved.

Unarmed Combat

Unarmed combat is used only as a last resort. The ninja will avoid unarmed combat to the best of his ability. It is within the nature of Ninjutsu to tactically set up the enemy so that he can be attacked and terminated with minimal struggle. The ninja has an assortment of weapons that he can use to terminate or combat his enemy. A well trained individual will use weapons to terminate or combat his enemy and under emergency conditions, he can utilize makeshift implements as lethal weapons. Unarmed combat is used only under emergency conditions when he can not utilize anything but his bare hands.

Effective Hand to Hand Combat can be a valuable tool to the ninja when he finds himself unarmed and threatened. The ninja concentrates and becomes very proficient in three fields of

Modern Day Ninjutsu 75

unarmed combat. The bare kill, which concentrates on quickly and ruthlessly attacking the enemy's vital targets causing rapid disablement or death. He also becomes proficient at lock and hold escapes so that he can quickly escape an enemy that is attempting to hold or arrest him. The ninja should also become proficient in disarming maneuvers so that he can escape and reverse the advantage of an armed enemy. He is well aware that his enemy is a dangerous adversary. He knows that killing an enemy with his bare hands is a difficult, strenuous and dangerous task. He is open to the possibility that the enemy may be a vicious, effective fighter that may defeat him in hand-to-hand combat. It is for those reasons that the ninja will make maximum use of the tactical advantage given by unexpected or surprise attacks and the technical and tactical advantages that can be provided by weapons.

The Bare Kill

Bare kill techniques consist of attacks on the enemy's vital targets. Attacks to the enemy's throat, temples and base of cerebellum are designed to kill. Attacks to the enemy's eyes, groin, knees, sides of neck, ears, kidneys and spine are designed to disable. The ninja may find it necessary to first weaken or disable his enemy before he can kill him.

All attacks must be executed with ruthless determination. It is the fighting fury that the ninja must possess, supplemented by effective, well executed attacks that can potentially kill the enemy. The ninja must be well aware that a fighter can absorb tremendous punishment. The fighter must attack powerfully and be capable of continuing his attack in a skillful, ruthless manner in order for him to destroy the enemy. The bare kill demands that the fighter quickly close in with the enemy and execute his attack. The fighter should be aware of the various angles of attack and defense. Likewise, he should be proficient at trapping the guard and exposing the enemy's vital targets. The fighter should be familiar with defensive maneuvers, especially jamming. Defense maneuvers are to be used just long enough to neutralize the enemy's attack.

In hand to hand combat, as in any form of combat, the ninja concentrates on the offense.

This chapter will provide the basics for unarmed combat.

Unarmed Kills

The following are simple, highly effective techniques that, when executed with fierce determination, can potentially kill.

Full Choke Hold

The full choke hold can be applied from a rear unexpected attack. The choke can also be applied on an injured enemy by forcing him to turn his back to you, then by manipulating his face backwards with either a hair pull or a nose gouge. The enemy's throat will then be vulnerable for the choke to be applied. When applying the full choke hold, be certain that the boney section of your forearm and wrist attack the enemy's throat. When applying the choke, be certain that the enemy's balance has been broken. The full choke hold correctly applied can result in a rapid death. The choke must be held for a minimum of 30 seconds. In the event that the enemy jams your choke attempt or succeeds in breaking your hold, you can retaliate. Quickly place your hands over the enemy's eyes and sharply pull his head backwards by applying pressure on his eyeballs. Brace the enemy's head against your abdomen as you force him backwards and towards the floor to a sitting position. Maintaining his head braced against your abdomen, sharply lean forward executing a neck crane technique. The neck crane can rapidly break the enemy's neck. If the enemy jammed or escaped your maneuver, be certain to inflict as much damage to his eyeballs as possible by raking his eyes. Continue a vicious attack.

Grip and Crush Attack to Adam's Apple

This attack is designed to crush the enemy's cartileges which make up his Adam's apple. Once the Adam's apple is crushed the enemy's throat will engorge with his own blood and his windpipe will close. The enemy will eventually fall unconscious and die shortly thereafter. When applying the technique, be certain to grip the enemy's throat high so that his Adam's apple is securely gripped in the attack. If the enemy attempts to jam your attack by lowering his chin, you can use your free hand to force his head back by utilizing a nose gouge or a sharp palm heel strike to the tip of his nose.

Modern Day Ninjutsu

Hook Punch to Temple

A well focused, powerful hook punch or barrage of hook punches to the temple can kill the enemy by rupturing the meningeral artery and causing massive hemorrhaging. The fist can be turned thumbside down in order to provide maximum penetration.

Knuckle Attack to Temple

The four knuckles or the middle knuckle can be used to attack the temple region. The knuckle attack allows for a more penetrating blow. The fighter must be aware that the knuckle attack can not be executed with as much force as a fist attack because of the resulting damage that can occur to his knuckles. This attack must be well focused and executed with sharp control.

Hook Forearm Punch to the Base of Skull

This attack is executed by smashing the inner part of the forearm against and through the enemy's base of skull. The blow resembles a rabbit punch except that the forearm is used as the striking weapon. This technique has the potential of breaking the enemy's neck. It can be used in an aggressive, continuous attack when the enemy turns and exposes his base of skull. The attack can also be used by positioning yourself to an angle to the outside of the enemy's guard.

Elbow Forearm Smash to Base of Skull

This technique can be executed from a position behind the enemy. It is executed by smashing the forearm elbow section against and through the enemy's base of skull. The attack has the potential to break the enemy's neck.

Stomp Kick Attack on a Fallen Enemy

If the enemy has been floored and is in a disabled or semi-disabled state a stomping attack can be used to terminate him. The heel of the foot can be forcefully stomped down and through the enemy's throat, neck, base of skull or temple region with lethal results.

The following are effective techniques that can be used to disable the enemy.

The Full Choke Hold is applied unexpectedly from behind. Break your opponent's balance and hold the choke for a minimum of 30 seconds.

Removing sentry using a two-man team. 1) Clasp enemy's throat and strike it 2) Break his balance 3) Second team member attacks.

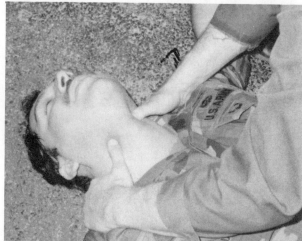

Throat grip and crush; if properly applied will kill an opponent.

Forearm Smash to base of skull is executed by smashing the forearm-elbow section against the enemy's base of the skull. This attack can break an opponent's neck.

Modern Day Ninjutsu

Spear Hand to Eyes
The enemy can be disabled by thrusting the rigid fingers into and through his eyeball. This technique can be used effectively from a close combat struggle as well as from a standard attack. The enemy's forward guard can be trapped to facilitate the attack.

Punch to Ear
A powerful hook punch or straight punch to the ear can be used to disable the enemy. The techniques can be used from a battery attack or from a standard attack.

Rigid Forearm Smash to Side of Neck
This technique is executed by smashing the inner section of the forearm against and through the enemy's side of the neck. The technique is executed in a semi-hooking manner. A powerful attack can result in unconsciousness.

Grip and Crush Groin
In a close struggle, the enemy's testicles can be grabbed and crushed. The testicles should be grabbed between the fingers and thumb to ensure a secure grip.

Front Kick Groin
The front kick can be used to attack the groin and disable the enemy. The attack is best used on a stable enemy from a surprise attack.

Roundhouse Kick to Groin
The roundhouse kick can be used to attack the enemy's groin. The kick should be modified slightly so that it raises upwards into and through the groin. The technique can be effectively used on an enemy that has assumed a fighting stance. An immediate attack should follow the technique.

Hook Punches to Kidney and Lower Spine Region
A series of powerful hook punches can be executed to the kidneys and lower spinal region. The attack can be used when the enemy turns away in a defensive manner or it can be used from an angular assault. A continuous attack should be maintained until the enemy collapses.

Side Kick Knee

The enemy can be disabled by sharp, forceful side kicks to the knees. This attack can be very useful against an enemy armed with a knife.

The following are simple compound maneuvers that can be used to help neutralize the enemy's offense, defense and counter offensive techniques while simultaneously facilitating the execution of your attack.

- A guarded rush can be used to pin the enemy to a wall, from this close in position, the enemy's groin can be grabbed and crushed or the enemy's head can be pulled downwards to expose the back of his neck to attack.

- A forward arm grab and trap can be used to stabilize the enemy while you execute a side kick to the knee or a front kick to the groin. A reverse straight or hooking punch can also be executed to the head with the aid of a forward arm trap.

- The enemy can be pinned against a wall off balance with a pushing grab. A forearm elbow smash can follow to the neck, face, throat, or head. Hook punches or knuckle attacks can also be used to the temple region.

- A loud, sharp scream executed just prior to your attack can momentarily stun the enemy.

Lock and Hold Escapes

The following are escapes from some common holds and some specialized locks. The ninja must be capable of quickly escaping from these locks and holds or his mission and life can be jeopardized. The principle of lock and hold escapes is to either jam the attempt and quickly attack or to react rapidly and execute the escape maneuvers before the lock or hold has time to be subduing or damaging.

Bearhugs

Bearhugs applied over the arms can be broken by applying groin grabs. The frontal bearhug applied under the arms can be broken by sharply attacking the enemy's eyes. The rear bearhug applied under the arms is the most difficult to escape. It can, however, be easily jammed by forcing your elbows and arms inward. The hold can be broken by a barrage of attacks on the enemy's exposed targets. Heel stomp kicks can be executed to the shins and insteps. The enemy's fingers can be grasped and torn outward. The enemy's face can receive jarring rearward headbutts. The fighter can step forward and lean his upper body backward to apply great pressure on the enemy's arms and grip, possibly breaking the hold. The fighter can trap the enemy's arm at the elbow with his own arm and execute a forward shoulder roll, arching his body so that it rolls over the enemy's face. On the floor, the enemy's groin can be attacked.

Choke Hold Escape

The choke hold should be jammed by tucking the chin downward and inward. The fighter's first concern is to jam or reduce the effectiveness of the choke by tucking and turning his chin. The second major concern is to regain balance. Once these two considerations are met, the fighter can execute a shoulder throw by firmly grasping the enemy's choking arm and violently pulling downward and across his body.

Collar Hold Escape

A strong hold on the collar or clothing can be broken with a forceful punch or palm thrust to the nose. An eye gouge is just as effective.

Arm Elbow Lock Escape

The outside arm elbow lock can be broken by raising the trapped hand with your free hand while thrusting your knee into the enemy's knee. The inside arm elbow lock can be broken by executing a sharp eye gouge forcing the enemy backward and to the floor.

Wrist Lock Escape

The wrist lock is best escaped by jamming the lock. With your free arm, smash down on the enemy's attacking arms. If the wrist lock cannot be totally stopped and a struggle begins, you can bite any exposed area on the enemy to help you escape. If an outside walk-along wrist lock is applied, you can escape it by pushing up on it with your free hand while simultaneously executing a knee strike to the enemy's knee. If an inside come-along wrist lock is applied, you can escape it by quickly apply-

Thrusting the rigid fingers into the eyeballs will quickly disable an assailant.

Forearm Smash to side of neck. Execute this attack by smashing the inner section of the forearm against the enemy's side of the neck in a semi-hooking manner.

In close-quarter struggles, a groin grab can be successfully employed.

A well focused, powerful blow to the Adam's apple can instantly disable an opponent.

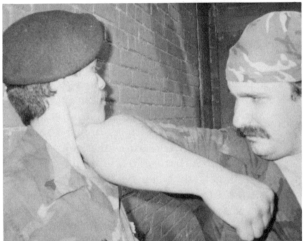

Delivering elbow to the Solar Plexus. This is one of the weak areas of the body.

ing an eye gouge that will force the enemy backward to the ground.

Disarming Maneuvers

The following are effective disarming maneuvers that can be used to neutralize and disarm an enemy. The ninja will avoid combatting an armed enemy with his bare hands whenever possible. The disarming maneuvers are to be used only under critical conditions. The principle of disarming is to act or react quickly to the threat by executing unexpected, effective attacks on an unprepared enemy. If an initial surprise attack can not be executed, the ninja may be forced to combat an armed enemy unarmed, or in the case of firearms, be forced to wait until the enemy makes a mistake offering him an opportunity to react. In either situation, the odds are against the unarmed person.

Disarming Maneuvers Against the Stick

The ninja is aware that a trained enemy armed with a stick has the advantage over the unarmed fighter. By quickly closing in and executing a trap and an attack, an enemy armed with a stick can be neutralized and disarmed. If the ninja is combatting an enemy armed with a stick, he must remain aware that his hands, wrists, forearm bones, elbows, head, face, neck, solar-plexus, collarbone, ribs, hips, groin, knees and shins are potential targets that can be fractured or injured. If the ninja must suffer a blow in order to close in, it is preferable that the blow be absorbed by a muscular padded section of the body. The muscular section of the forearm can be used to deflect a blow. Care should be taken not to receive a solid impact on the forearm, as it can result in a disabled arm. The fighter must move very quickly and agile when confronting the stick fighter. He should be proficient at quickly ducking and weaving under and to the sides of the swinging stick. The fighter must concentrate on getting in close to the stick fighter in order to neutralize him. As soon as the fighter closes in with the enemy, he can trap the arm with the stick and execute his attack.

Disarming Maneuvers Against the Knife

The ninja is well aware that an enemy armed with a knife has the advantage. By quickly closing in and executing an effective, ruthless, unexpected attack, he can neutralize and disarm the knife fighter. The unarmed fighter must make every effort to keep the knife fighter in an unstable defensive state. He must be trained to take advantage of any hesitation or insecurity that can be installed in the enemy.

If the unarmed ninja must combat the knife fighter, he can move towards the knife fighter's empty-handed side. This maneuver can offer the unarmed fighter a momentary tactical advantage from which forceful, penetrating side kicks to the enemy's knees can be executed. The ninja can execute a sudden dash to close the gap and to trap the enemy's knife arm while simultaneously executing an attack. Defensive maneuvers against knife attacks consist of deflecting or parrying the attacking arm followed immediately by a trap and counter attack. Biting and groin grabs can be used in close quarter struggles against the knife fighter.

Disarming Maneuvers Against the Handgun

An enemy armed with a handgun can be neutralized if his intentions are to hold you prisoner and if he is not expecting an attack. The best way to neutralize an enemy armed with a handgun is to attack him while he is unprepared to shoot. The ninja must concentrate on keeping his body out of the line of fire and on controlling and disarming the enemy of his weapon. The handgun can be trapped and controlled while an attack is made on the enemy's vital targets. The handgun must then be given total attention and torn from the enemy's grip. An alternate plan is to commit oneself totally towards tearing the handgun out of the enemy's hand. In either attack plan, the ninja must gain superior leverage over the enemy and always maintain himself out of the line of fire. The disarming techniques are as follows: The handgun can be torn out of the enemy's hand by grasping the armed hand's wrist with one hand and by pushing the gun's barrel backward towards the outside or the inside of the enemy's forearm. Either technique will apply great pressure on the enemy's fingers and wrist. An unsuspecting enemy can be disarmed of his handgun if he is close to you, whether he holds it to your front, rear, body or head.

If he is to your front and holds the gun aiming it at your body or head, you can quickly pivot your body and head out of

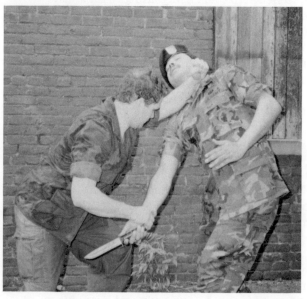

To disarm a knife fighter the Ninja must execute a quick, ruthless, unexpected attack.

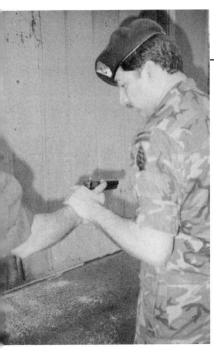

Handgun disarming. Grasp the enemy's wrist with one hand and push the gun's barrel backwards toward the inside of his forearm.

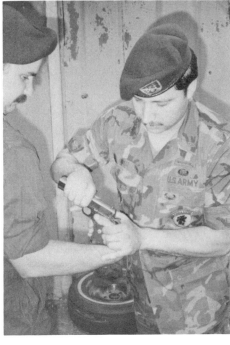

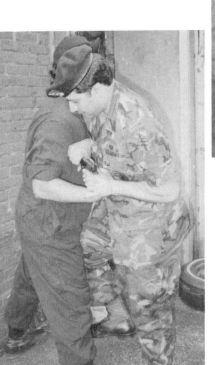

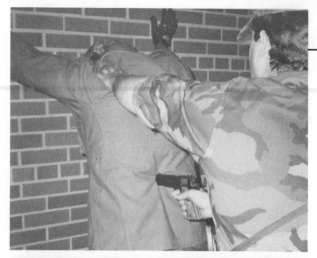

Advanced pistol disarming. If enemy aims his gun at your back, quickly pivot your body to the outside of his gun while executing a parry. Deliver forearm smash to back of neck; follow immediately with a trap on the gun's arm. Disarm by forcing the gun **backwards** and kick the groin forcefully.

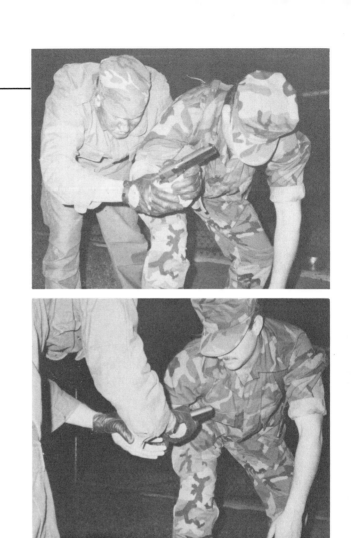

the line of fire as you simultaneously step forward and execute the disarming techniques. If he holds it pressed against your head, you can parry his arm away as you move your head out of the line of fire. You must then continue to move as you execute a trap and a disarming technique. If the enemy holds the gun in his rear hand and keeps his unarmed hand forward, you can quickly move toward his unarmed side and execute an initial strike to an exposed target. You will be temporarily out of the line of fire. Immediately maneuver around the rear of the enemy or force him to turn and get control of the handgun by executing a disarming technique. If the enemy holds the gun aiming at your backside, you can quickly pivot your body to the outside of the enemy's armed side while executing a parry. Follow immediately by executing a trap on the handgun's arm and execute a disarming technique. If the enemy holds the gun to his rear and keeps his face hand forward, you can spin to the outside of his forward hand and execute an initial attack. You then quickly maneuver behind the enemy or force him to turn and concentrate on disarming.

The unarmed combat techniques and maneuvers presented are simple and potentially effective. The ninja must be aware that they must be executed with determination and ruthlessness in an unexpected manner. They can offer the unarmed fighter an excellent chance of survival, but there is no guarantee. The level of proficiency on the part of the ninja will be matched against the degree of proficiency held by his enemy. The results of the encounter can never be assured. It is for these reasons that the ninja will make maximum use of weapons and if he is forced to combat unarmed, he will arm himself as soon as possible. *It should be noted that different schools and masters have developed their own specialized maneuvers. When evaluating specialized maneuvers the questions of practicality and potentials should be raised.*

Weaponry

This chapter is designed to familiarize the reader with various weapons that can be used to combat the enemy. The chapter will cover traditional, modern, specialized and miscellaneous weapons that have practical value for modern day combat survival. Weapons can offer the ninja a practical as well as psychological advantage in combat. The practical aspects are the potential effects of the weapon. Aside from being a fighting instrument, a weapon can be a source of motivation, confidence and determination. Possession and expertise with a weapon can reinforce the ninja's confidence and fighting spirit. Weapons should be an important consideration when preparing to engage and combat the enemy.

The following are some tactics that can be used to maximize the potential effects of weapons.

Practicality

The ninja must choose weapons that are suited for his mission. The purpose of the mission, the range in which the enemy is to be engaged, the method by which to close in and engage the enemy, the enemy's state of readiness or lack of it, the enemy's means of defense, the enemy's training, and the number of enemies expected to be engaged should determine the weapons that will be utilized. The ninja should plan to carry reserve or back-up weapons to meet emergency situations.

Ninjas should consider the following weapons as standard weapons for modern day combat survival: the rifle, the pistol, the knife and the bare hands. He should become proficient with the mentioned weapons because they are practical and they suit the demands of varied missions and situations.

Expertise

Before attempting to utilize a weapon in a mission, the ninja should become an expert with his weapon. He should become proficient with the combat offensive and defensive uses of his weapons and should feel very comfortable with them. The weapon should be considered and treated as an extension of the hand.

Effectiveness of Weapon

The ninja must be fully aware of the advantages and disadvantages of his weapons. He should also know the advantages and disadvantages of the enemy's weapons. The weapon's effective range, its possible malfunctions, the potential injuries and damages it can cause, and its potential for use in instinctive fighting are all considerations that the ninja must be fully aware of.

Concealment and Carry

The method of carry and the means by which to conceal the weapon are important factors. The weapon should be carried in a comfortable, safe manner which will permit its immediate

Modern Day Ninjutsu 95

instinctive use. If the weapon is concealed, it must be concealed in such a manner so that the ninja can quickly draw and utilize it without enduring time consuming procedures.

The following are some methods that can be used to conceal weapons that will not interfere with their immediate use:
- holsters properly designed to carry and conceal the weapon
- A long overcoat can be worn to cover and conceal weapons.
- Weapons can be carried in hand covered and concealed by a hand held jacket or a jacket tossed over the arm.

Surprise Attacks

Whenever possible, the ninja will attack in an unexpected manner. He will always attempt to destroy the enemy when and where he is least expecting or prepared for an attack.

Ruthlessness

The ninja must be fully prepared to carry out his mission. He must be psychologically prepared and conditioned to fight in a ruthless manner. He must be motivated and determined to dispose of his enemies. An aggressive, ruthless attitude is a requirement for combat survival.

Intimidating Effects

The ninja should be aware of any intimidating effects his weapons carry. The enemy can be shocked and demoralized by the sight and effects of certain weapons.

The following are weapons that can be used to combat and destroy the enemy. It is suggested that the ninja familiarize himself with the various weapons and their combat uses. He should then choose and master a few weapons that he finds most comfortable. The ninja should include the standard weapons, previously mentioned, in his arsenal. He should attempt to choose weapons that can serve him under varying circumstances.

WEAPONS

Rifle

The rifle can be a very valuable and effective weapon for the modern day ninja. The rifle can be used basically for two purposes: long range firing and as an assault weapon.

If the ninja plans to employ his rifle for a long range hit, he must choose a rifle that is very accurate and he must become an expert shooter with his weapon. The following are some rifles that have a reputation or the potential for long range firing: The Remington 700 in either .308 or .223 caliber, the M14 (accurized), the Winchester model 70 .30-06, and the M16A1.

The mentioned rifles can be effectively employed at varying ranges. For example, the M16A1 can be effectively used without a scope for a range of 460 meters. The Remington bolt action .308 with a X12 scope can be effectively used at a range of over 1000 meters. The user must determine at what ranges he will be firing and choose his weapon and accessories accordingly.

To employ the rifle effectively in long range shooting, the ninja should be proficient in field crafts or urban maneuvers, cover and concealment, basic rifle maintenance, communications—if applicable—and escape and evasion, as well as being an expert shooter from the various shooting positions. Telescopic sights can be mounted on the rifle and a starlight or an infrared scope should be used during periods of low visibility. If feasible, the use of a noise-suppressor (silencer) should be considered. Laser scopes can be considered.

If the ninja plans to use his rifle as an assault weapon, he should choose a semi-automatic or semi-full automatic rifle. The M16A1, carbine, M45 known as the Swedish K, the AR15 and the UZI can be used as effective assault weapons. For assault purposes, the ninja should become an expert at instinctive shooting both from the hip and from the shoulder position.

Instinctive shooting is executed by locking the rifle at the hip firing position or the shoulder firing position. When the ninja has cause to shoot, he pivots and turns his body in the direction of fire and fires maintaining his weapon in the locked position. In this manner, the ninja will instinctively shoot at the target which he is facing. The preferred targets for a kill are the head and the heart. Strive to hit those targets in training and in combat.

The Modern Day Ninja should become acquainted with all types of weapons. 1) Ninja armed with short sword, K-Bar knife, Colt .45 pistol, and silenced M16A1 automatic rifle 2) Winchester .308 rifle with X9 scope and noise suppressor 3) Camouflaged Ninja with M16 rifle.

Handgun

The handgun can be a very valuable and effective close quarter weapon to the modern day ninja. For the purpose of close quarter combat shooting, he should utilize the heaviest caliber he can effectively handle. The .45, .44 magnum, and .357 magnum are powerful handguns. The ninja should become proficient at target shooting and instinctive shooting. Instinctive shooting or point shooting is executed by locking the wrist and the elbow and positioning the arm by the side. When the ninja has cause to shoot, he pivots and turns his body in the direction of the target. The rigid arm is brought up centered on the chest. The free arm can be held to the side for balance or it can be crossed in front of the chest to offer some protection. The ninja will instinctively shoot at the target he faces. The preferred kill targets are the head and the heart. The modern day ninja should condition himself to instinctively fire two or three round bursts at his targets. If the target is only one enemy, then the whole pistol or revolver can be emptied on the enemy. The ninja should have extra magazines or auto load rounds to quickly re-load his weapon. Hollow-point bullets can be used to cause maximum damage. If available, teflon coated bullets can be used to penetrate bullet proof vests or body armor.

If the handgun is to be silenced, then a .22 caliber can be used. The .22 caliber can be effectively silenced and will produce a kill if the enemy's brain or heart is destroyed.

The ninja can carry a small handgun as a back-up or reserve weapon. The small handguns can be concealed on the wrist, ankle, the small of the back, or between the shoulder blades slightly below the neckline. The following are small handguns that can be used as reserve weapons to kill the enemy. The Remington .41 caliber Derringer, the .38 caliber Smith & Wesson Derringer, the Colt Astra Cub Automatic in .25 or .22 caliber and the Walther PPK .380 caliber. The small handguns will cause maximum damage if fired through the eye sockets into the brain, through the base of the cerebellum into the brain, through the open mouth to sever the spinal chord, at a point just below the ear to sever the spinal chord or through the temple severing the meningeral artery. Handguns should be ideally used at very close quarters. A specialized weapon that can be of use is a .22 cal. or .25 cal. single shot pengun.

Handgun Stress Shooting

The Modern Day Ninja should always attempt to have cover and concealment when engaged in a shoot out and whenever possible, he should set up his enemy so that he can gain the tactical edge of surprise.

In the event that you are denied your strategy, you must be prepared to react to a confrontation in a proficient manner. As mentioned before, you should be skilled at quickly drawing and executing the instinctive or point shooting techniques. The following is a maneuver that can be used in conjunction with point shooting as a defensive reaction. The maneuver is termed Drop Shooting and it is ideally suited for situations when you are caught momentarily at a disadvantage in a surrounding offering no cover.

Drop Shooting

For this maneuver it is required that you become proficient at quickly dropping to the ground, on your unarmed side. For example, if you shoot right-handed then you must drop on your left side. It is required that you drop so that your feet point at the enemy. Your head must be kept low to the ground and your knees are drawn up towards your body, feet together. It is preferable that your handgun be drawn at the time of the drop but in emergencies the handgun can be drawn from the ground position. The handgun should be rested on top of your top foot, the barrel slightly protruding past the shoe. In this manner you can shoot at a close enemy with accuracy. The enemy which adopts the combat crouch will be vulnerable to your shots while the primary targets facing the enemy will be the soles of your shoes and your buttocks.

This technique can be of use against enemies armed with firearms or with close quarter weapons. In the event that your enemy was charging you with a close combat weapon, for example a knife, you could shoot him as he charges, meet and stop his charge with your feet and continue to shoot him.

Experts at this technique can drop and shoot fast enough so

Handguns are effective close-quarter weapons. Photos show correct Weaver stance.

Drop shooting. To execute this maneuver drop quickly to the ground, while at the same time drawing your gun, rest your gun on your top foot and shoot upwards.

as to shoot and neutralize an armed enemy while sustaining no injury or wounds only to their lower extremeties.

When executing the drop shooting maneuver it is best to catch the enemy temporarily off-guard. The maneuver itself is an unexpected one and if executed rapidly it may save your life.

Knife

The knife can be a valuable and effective weapon for the ninja. The KABAR U.S.M.C. combat knife is an excellent weapon that can be used as a primary weapon or as a reserve weapon to engage and kill the enemy. If you require a smaller easily concealable knife, the Gerber Mark I would be an excellent weapon. A more easily concealable weapon is a folding knife. The folding knife should have a lock, be razor sharp and be easily opened with one hand. The small folding knife may be limited in its cutting potentials, however, short thrusts to the eyes and neck can be used effectively. The folding knife is best used from surprise, unexpected attacks to the enemy's eyes and neck. It is to be considered as a reserve or emergency weapon. Pen knives are also considered reserve or backup weapons. They can be of value under specialized conditions.

The ninja can use his knife to silently kill an enemy. From the rear, he would clasp the enemy's mouth and nose and pull back sharply, breaking the enemy's balance. The knife would then be used to cut the enemy's neck and throat. The ninja can use his knife to kill the enemy from a forward walking maneuver. He would walk in the direction of the enemy, keeping his knife in the ice pick grip pressed and concealed against his outside forearm. If the enemy is to the left, then the knife is held in the right hand and vice versa. When the ninja closed in with the enemy, he would turn towards him and using an outward circular attack, cut the enemy's neck and throat.

If the ninja must combat the enemy, then he would immediately cut the enemy's fingers, back of the hands or forearm to force the enemy to withdraw his guard and retreat to a defensive attitude. The ninja will as quickly as possible cut the enemy's neck and throat or stab the enemy in the solar plexus. If he stabbed the enemy, the knife can be twisted and pumped to cause massive hemorrhaging.

The enemy's neck, containing the jugular veins and the carotid arteries, is the preferred kill target. The subclavian artery located under the clavical bone; the ulner artery located on the thumb side of the wrist; the brachial artery located on the little finger's side of the inside of the elbow; the heart; and the femoral artery located on the inner side of the thigh are all potential kill targets that can be attacked as they present themselves.

The ninja must become very fast and proficient with his knife. Inward, outward, circular, and downward cuts can be used accompanied by thrusts and quick slashes to overcome and kill the enemy. Defensive maneuvers consist of avoiding the attack and executing a cut to the attacking limb or an avoidance maneuver followed by a quick counter attack.

In combat, the ninja's stance should be one that allows him quick, reflexive movement. The guard should be close to the body. The knife can be held in the forward or rear hand in the sabre or ice pick grip.

A word on knife throwing: The ninja will never throw his knife at an enemy. The odds against the knife hitting the enemy and causing a disabling or lethal effect are great. The only circumstance in which the ninja would throw his knife is when the enemy is fairly close—approximately 10 or 15 feet—and he is armed with a firearm or a long-range weapon such as a crossbow. Under these circumstances, he can quickly throw his knife at the enemy's neck or upper chest to cause a temporary distraction which may permit him to close in, disarm and destroy the enemy. It will not matter if the knife penetrates or not. The most important factor is that it be thrown quickly and forcefully to cause the enemy to guard against it. This will in turn cause a momentary hesitation and therefore, offer the ninja a chance.

Shotgun

The shotgun can be a very lethal close quarter weapon. It can be used effectively from distances of 10 to 25 yards. The ninja should carry the most powerful shotgun he can effectively handle. A 10 gauge or 12 gauge shotgun loaded with buckshot or .25"/6mm steel ball bearings can be an excellent choice. If

A sharpened U.S. Marine Corps (K-Bar) Combat knife is an excellent weapon.

Concealed knife grip.

One of the correct knife stances.

Outward circular cut to Carotid artery. This maneuver is executed from the concealed knife grip.

Thrust to heart through the Solar Plexus.

Knife thrust to neck.

Blocking blunt instrument from the kinfe-in-rear stance.

Knife thrust to groin or Femoral artery.

To stop a knife attack 1) Parry the knife with the left hand while cutting at the Ulnar artery with the right hand 2) Continue attack by thrusting knife to the opponent's throat area.

Modern Day Ninjutsu

the shotgun is to be employed at very close quarters, the barrel and stock can be sawed down. A sawed off pump action or double barrel 12 gauge or 10 gauge shotgun loaded with buckshot can be concealed with a long overcoat and can be used at close quarters with lethal effects. The ninja can choose a double barrel or a pump action shotgun. The pump action shotguns can have their plug removed and can hold 5 cartridges. The preferred kill target is the head.

A specialized weapon that can be of value is the shotgun pistol. The weapon is concealable and can be used effectively at close range.

Crossbow

The crossbow can be used to silently kill the enemy at a distance of 5 to 30 yards. The following are two crossbows that can be effectively utilized.

The CMX COMMANDO Crossbow comes with a 100 lbs., 125 lbs., 150 lbs., or 175 lbs. pull. A scope can be mounted on it for better accuracy.

The WCX WILDCAT Crossbow comes with a 150 lbs. pull and can also be mounted with a scope.

The BUSHNELL 74-1403 X4 (32mm) scope is a scope designed to be mounted on the mentioned crossbows.

The crossbows should nock a 3 point broadhead arrowpoint with an aluminum or graphite shaft. The standard deer hunting or bear hunting fletches will suffice. The broadheads should be razor sharp to cause maximum damage.

The ninja should become very proficient with his crossbow, realizing the effects of wind or rain before he attempts to use the crossbow in combat. The preferred kill targets are the neck and the heart. The ninja can choose to poison or contaminate his arrowpoints.

Bows and Arrows

The bow and arrows can be used to silently kill an enemy from 10 to 25 yards. Ninjas can choose from either the compound bows or the recurve bows. Choose a bow with the heaviest pound draw you can handle effectively. The minimum pound draw should be 40 lbs. The compound bows will permit a heavier pound draw; however, compound bows need more care than regular recurve bows. Use very sharp 4 point broadheads on aluminum or graphite shafts with standard deer or bear hunting fletches. The bows can be fitted with sights for better accuracy. Several sight pins can be fitted for varying distances, although for close quarter shooting, they are not required. The ninja should become very proficient with his bow and be capable of instinctive shooting with accuracy as well as nocking his arrows quickly and correctly. You can choose to poison or contaminate the arrowheads. The preferred kill targets are the heart and neck.

Tranquilizer Guns or Bio-Innoculator Weapons

Tranquilizer guns can be loaded with poison to silently kill the enemy. The ninja should become proficient with the weapon before he uses it in combat. The tranquilizer gun is a specialized weapon that can be used in special missions for a silent kill.

Garrote

The garrote is a specialized weapon that can kill the enemy silently. Makeshift garrotes can be made from fish lines or boot laces and two pieces of wood. A necktie can also be converted into a makeshift garrote. The ninja can silently kill his enemy by closing in from behind and looping the garrote around the enemy's neck. He would then sharply pull back and downwards while simultaneously driving a knee into the enemy's back to break his balance. The garrote would then be applied until the enemy dies.

A belt can be used in a similar manner to break the enemy's back. The belt would be looped over the enemy's neck, then sharply pulled backward and down while the knee applies pressure forward on the small of the back.

Machete

The machete can be a fearsome, lethal weapon in close quarters. The ninja should become proficient at quickly wielding it in a circular manner to attack and kill the enemy. From an unexpected attack, the machete can be used to attack and sever the neck, back of the neck, throat and face. In combat, the enemy's fingers, hands, wrists, arms, knees, abdomen, clavical, neck, throat and face can be attacked with effect. The preferred killing target is the neck.

Handaxe

The handaxe can be used with lethal effects at close quarters. It can be wielded similar to the machete to attack the various targets. From a rear, unexpected attack, the handaxe can be used to kill the enemy by smashing the blunt edge of the axe through the enemy's back of the neck. The preferred kill targets are the base of the cerebellum or the temples.

Teargas

A very potent teargas or liquid can be used to momentarily incapacitate the enemy. The disadvantage of the gas is that in an open area with wind or rain, the gas can be rendered ineffective. The disadvantage of the liquid is that if it is contained in an oil base, it can take a second or two to take effect. If the enemy quickly wipes his face, he can weaken the effects. The disadvantages of both the gas and the liquid is that a highly motivated enemy may not be stopped. Teargas can be used as a momentary distraction followed immediately by an attack. Formaldehyde concealed in a nasal spray is an effective deterrent. Teargas pens should be considered as specialized weapons. It should be noted that teargas containers can be loaded with more lethal substances.

Wallet and Belt Buckle Knives

Wallet and belt buckle knives are small hand-held "push dagger" type knives. They can be used as emergency weapons. The knives are best used in an unexpected attack against the throat, neck or eyes.

Nunchaku

The nunchaku is a traditional oriental martial arts weapon. If the ninja chooses to use the nunchaku, he should become proficient with swinging it in circles. This is important so that he loses his fear of the weapon. After the ninja is comfortable with the nunchaku, he should concentrate on developing his attacks. The nunchaku is to be employed with great speed. The ninja should learn how to attack and then maintain control of his weapon after impact. The striking stick of the nunchaku has a tendency to snap back and hit the operator in the face or hand if it is not properly pulled back and controlled. Learn how to instinctively turn and pull the nunchaku after impact in order to keep it in control and ready for a continuous barrage of blows. It is not necessary to become proficient at the spectacular hand switching and twirling maneuvers with the nunchaku for combat. The maneuvers, however, can be used to help the operator become more familiar with his weapon. The targets are the hands, forearm bones, knees, face, head, clavical, elbows, shoulder bones, wrists and shins. The preferred killing targets are the temples and base of skull. The nunchaku can, aside from being swung, be used as a thrusting weapon to the throat, face and solar plexus. It can also be used to choke the enemy. The ninja should attack in a ruthless, unexpected manner with his weapon because it is possible to check, deflect, catch and consequently disarm the nunchaku wielding enemy.

Sai

The sai is a traditional oriental martial arts weapon. The tips of the sai should be filed to a sharp point. The weapon can be used as a blocking, trapping and deflecting instrument against other nonprojectile weapons. The sai is used offensively by striking with its cylinder-like end or by thrusting the sharp tip through the enemy. It can be inversed and the end of the handle can be used to strike. The targets include the eyes, face, head, neck, throat, wrists, ribcage, groin and abdomen. The sai can be used to kill by thrusting the sharp end through the eyeball into the brain or through the throat or through the solar plexus to the heart or spine. The sai can also be used to thrust through the base of skull to the brain, or thrust through a point just below the ear to sever the spinal cord. The preferred kill targets are the brain through the eye socket and the throat.

Yawara

The yawara is a weapon of the traditional oriental martial arts. It is used by driving its protruding ends into the eyes, throat, neck, base of skull and temple. The preferred kill target is the temple.

Short Sword

The short swords which I refer to in this section are the

Japanese wakizashi and the Japanese NINJA-TŌ. The wakizashi is the traditional short sword of the Japanese samurai. The NINJA-TŌ is the traditional sword of Ninjutsu. In combat, the NINJA-TŌ has the advantage that it can be wielded with one or two hands. The wakizashi is a one handed weapon.

The ninja's weapon should be razor sharp and of a sturdy manufacture. The handle should provide an excellent grip. The user should become proficient at quickly drawing his weapon and in a continuous motion, executing an attack (*aido*). The ninja should also become skillful at offensive and defensive maneuvers (*kenjitsu*). A short sword wielded efficiently can be a very dangerous and effective close quarter weapon. The ninja should concentrate on utilizing his sword with great speed and accuracy. The targets for cuts include the face, neck, throat, hands, wrists, biceps, the area surrounding the top and sides of the knees, back of the neck, behind the knees and the achilles tendon. Thrusts can be delivered to the throat or abdomen.

The short sword can be a very valuable and effective weapon in close quarters. In a limited space such as a room or passageway, an expert with a short sword can engage and destroy multiple adversaries providing they do not have firearms and react professionally.

Kama

The Kama is a scyth-like weapon. It can be used to hack at the enemy or to entrap him. The preferred target is the neck-throat region. It can be used effectively to cut any of the enemy's vital targets such as the subclavian artery, groin or torso.

Bakuhatsugama

This weapon consists of a Kama with a weighted chain attached to the handle. The chain can be used to strike the enemy or to entrap him or his weapon while the blade is brought into use.

Kusarifundo or Manriki Kusari

This weapon is a chain weighted at both ends. It can be used to strike or entrap the enemy. The kusarifundo can be used to choke the enemy.

Other useful weapons include 1) Sawed-off 12 gauge shotgun 2) Crossbow 3) Compound bow.

1 & 2) A garrote is a frightful silent weapon. Here Ninja shows how it is used. Note leg entrapment; heels dug into groin.

3) Using the Sai to trap and thrust to throat. The Sai is a traditional Oriental martial arts weapon.

4 & 5) A Ninja tactic was to climb a tree and throw his sword at a passing enemy.

Specialized Kama with spiked knuckle guard.

Ninja disposes of enemy using two Kamas—One around the neck and the other around groin.

Ninja armed with a Bakuhatsugama. This weapon consists of a Kama with a weighted chain attached to the handle.

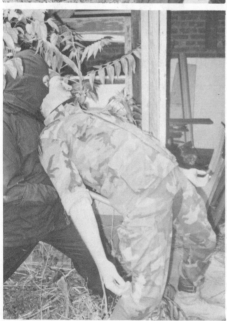

Manriki-Kusari chain attack. Using this fighting technique a short chain weighted at both ends becomes a potent weapon.

Steel Fan

This is a peculiar makeshift weapon. The fan has steel arms and can be used to block or parry shurikens or thrown knives. The steel fan can also be used to strike the enemy's vital points; particularly the eyes or throat via a thrust. The steel fan can also be used as a weapon for close combat utilizing parrying, slashing and thrusting motions.

Shurikens

Shurikens or throwing stars have been identified as peculiar to ninjutsu. In fact they are not. African cannibal tribes such as the Asandas have used throwing implements which look like a cross between shurikens and knives to capture and kill their prey.

Shurikens are essentially used as disabling weapons. They can be thrown in a variety of manners, the most powerful being the overhand throw. Heavy clothing may stop or diminish the effects of a thrown shuriken and as such the preferred target is the face-neck region. If the enemy is ill-clothed the shurikens can be thrown at the torso or limbs to disable. Some modern shurikens consist of four points and are constructed of surgical steel. Thrown with force they have been known to penetrate and pass through the door of an automobile. These shurikens can cause severe injury or death if accurately thrown. Shuriken tips can be poisoned for more lethal results.

The shurikens can be thrown individually or, by staggering them, you can throw more than one at a time. They will fan out when thrown side-arm. This method may be used when confronting more than one enemy which are positioned close together. If you throw more than one shuriken overhand they will spread out vertically and slightly fan outwards. The fan spreads can be modified depending on the particular style used.

Shurikens can also be used as in-close weapons. Hand held they can be used to cut and tear the enemy's flesh. With force the carotid, brachial, ulner, and meningerial arteries can be severed which may lead to death.

Tetsu-Bishi (*Caltrops*)

Caltrops are weapons which can be used to slow down an enemy or to disable him. Good caltrops will penetrate the average shoe depending on the speed and weight of the enemy. Poisoned and contaminated caltrops such as a caltrop coated with blood and dung can add to the disabling effects. In close combat caltrops can be thrown at the enemy's face to temporarily disable him. Caltrops can also be placed next to tires to cause a flat.

Tekagi

The tekagi is primarily a climbing device. It is used to scale walls. It can, however, be used to slash and tear the enemy's face. Legend has it that they can be used to block, parry or catch the enemy's blades. This is unlikely, since good blades can cut through steel. The ninja that attempts to catch a samurai's katana relying on the tekagi will more than likely lose his hands and his life.

Nekade

The Nekade are metal tips which are attached to the fingers. They can be used as an aid to climbing. They can also be used to tear the enemy's flesh. The Nekade can be used effectively against the enemy's face, particularly the eyes.

A makeshift Nekade can be made from a can's tab. By inserting the finger through the loop and positioning the flat edge against the fingers, a can tab can be effectively used to tear and lacerate the enemy's face and eyes.

Blowguns

Blowguns can be used in conjunction with poisoned needles to disable or kill the enemy. Blowguns come in many sizes from a long of three feet to a short of six inches disguised as a pen. In modern days a biological-innoculator dart pistol or rifle can be used to sedate or kill the enemy. The poisons curare or the venom from the Australian tiger snake, one of the most poisonous snakes on earth, can be used.

A specialized weapon which is effective at very close range is the miniature blowgun. This blowgun is completely concealed in the mouth until it is brought into use. A shortened drinking straw can become a makeshift miniature blowgun.

Night Stick

The night stick can be used to attack the bony areas of the

Steel fan. This weapon is used to block or parry Shurikens or knives.

Ninjas were adept at throwing Shurikens (throwing stars) and dirks. These come in many different shapes and sizes, including folding and armor piercing.

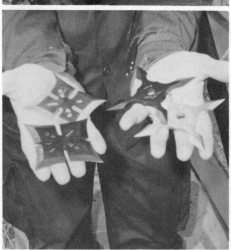

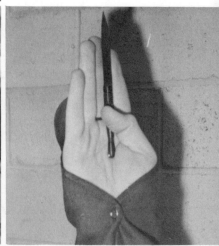

Tetsu-Bishi or caltrops are used for a variety of jobs including 1) Slowing down an enemy in pursuit 2) Disabling vehicles by causing flat tires and 3) as hand-held weapons in close combat.

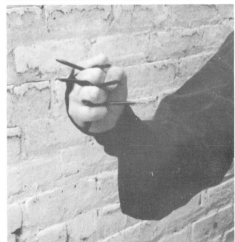

Common blowguns come in many sizes from three feet to six inches disguised as a pen.

Blowgun used in emergency! 1) As enemy prepares to draw his sword 2) Ninja strikes with blowgun to temple 3) Crouches and blows out dart to 4) Enemy's eye.

Regular and special modified bladed Tonfa.

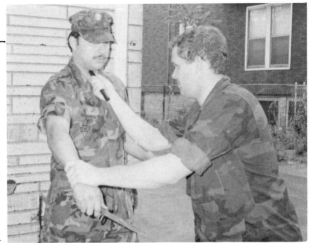

The Kyoga is a handle which when swung, will extend a weighted spring chain. Here it is used in the closed position to 1) Thrust to throat and 2) Groin; and in the open position to 3) Strike temple.

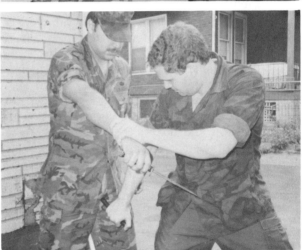

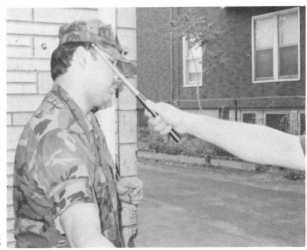

enemy as well as the throat, groin and solar plexus. It can also be used to choke the enemy.

Kyoga
The Kyoga or steel cobra is a handle which when swung will extend a metal protrusion or weighted spring chain. It is used to crush the enemy's vital points.

Shoge
This weapon consists of a knife and a circular edge. From the handle protrudes a long chain with a circular weighted end. The Shoge can be used to entrap, choke or to thrust and cut the enemy.

A specialized weapon that the Modern Day Ninja-agent has in his arsenal is the mini-grenade or the "button grenade." This is a mini anti-personnel grenade that is disguised as an overcoat button and detonates on impact. It consists of a button-shaped casing which incorporates two small vials, one containing red phosphorus and the other containing potassium chlorate. The button is attached to the overcoat. When required, the ninja removes and throws the button grenade at the floor next to the wall or enemy. On impact the two vials break and when the two chemicals mix, a spontaneous explosion occurs.

Potassium chlorate is almost as strong as T.N.T. and when it explodes it will produce a kill zone and a concussion zone. The ninja agent must be an adept chemist in order to prepare the button grenade and for him to know the kill, concussion and safe ranges when utilizing the button grenade.

Makeshift Weapons
The following are common implements that can be converted into lethal weapons under emergency situations. The makeshift weapons should be employed with great speed and an aggressive, ruthless attitude. The ninja should become very familiar with the makeshift weapons since they require little technical skill and they can be very valuable under emergency conditions.

Pen, Pencil and Key
The pen, pencil or key can be used to gouge the enemy's eyes. The pen can also be used to crush the enemy's trachea.

Ashtray
An ashtray of sturdy manufacture can be used to smash the enemy's nose, throat, base of skull, wrists and temples.

Pin
A pin can easily be concealed in a jacket's inner pocket or in various other areas. The pin can be used to attack the enemy's eye. The pin should have an adequate head for gripping. The head can be taped for a better grip. A large hat pin is preferable over the smaller versions.

Beer Mug, Bottle and Pitcher
A beer mug, bottle or heavy pitcher can be smashed against the enemy's face. The jagged edges of the remains can be used to attack the enemy's face and eyes.

Belt
A leather belt can be wrapped around the hand's knuckles to reinforce the force of a punch. The enemy's temples, jaw, chin, nose, neck, solar plexus, kidneys and ribcage can be attacked effectively with the aid of the belt.

Chair
A chair can be used to defend against a knife, chain or stick. By rushing the enemy, the ninja can overpower him and attempt to neutralize the weapon or execute low kicks to the groin.

Lock and Cord or Sock and Lock
A cord or a sock can be tied to a lock to make an effective makeshift weapon. The lock would be swung at the enemy's head, face or hands.

Screwdriver or Ice Pick
A screwdriver or an ice pick can be driven through the enemy's eye, temple or base of skull.

Straight Razor or Scissors
A straight razor or scissor can be used to cut the enemy's eye, face, neck and throat. The scissors can also be thrust through the eye or throat.

Pipes and Clubs
A length of pipe, baseball bat or length of solid wood can be used to batter the enemy's neck, base of skull, wrists, knees,

The spear in application against enemy.

Shoge demonstrated by Ninja is used to entrap, choke, thrust and cut the enemy.

The use of everyday items as weapons is a traditional Ninja specialty; here a fighter uses an ice pick to attack his enemy.

Spring loaded bolt shoots broadhead from pipe base strapped to arm. This weapon can be concealed beneath a jacket.

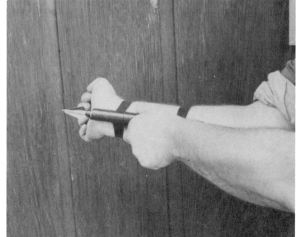

skull and side of head. The preferred kill targets are the base of skull and temples.

This chapter covered the various weapons that can be of use to the ninja. The fighter should treat the weapon as an extension of the hand or self. In combat, a weapon is a part of the body and is utilized in a very natural and comfortable manner. The following are two psychological approaches that are used when engaging an enemy armed or unarmed. The primary difference between the two approaches is whether to concentrate on tactics or techniques.

When engaging an enemy from an unexpected attack, the ninja should concentrate on his attack technique having already tactically set up the enemy. For example, if the ninja were going to shoot an enemy from an undetected, concealed position, he should concentrate fully on his shooting technique since he is already in a tactically advantageous position.

If the ninja is engaged in combat with the enemy, then he should concentrate on gaining a tactical advantage and depend on his trained reflexes to execute the appropriate offensive and defensive maneuvers. For example, if the ninja is in combat against an adversary wielding a baseball bat while he is armed with a knife, then he should concentrate on getting in close to his enemy—a tactically advantageous position for the ninja—where the enemy's bat can be easily neutralized; while the knife can be employed to cause serious injuries or death. The ninja would concentrate on tactics and depend on his conditioned reflexes to execute the avoidance, rushing, trapping and attacking maneuvers that are required to terminate the enemy.

Assassination

Assassination is the deliberate pre-planned killing of a human being. The ninja that specializes in the art of assassination must prepare himself for the stress and difficulties that he can encounter. A rule to be followed is to maintain the plan of action as simple as possible. Auto hypnosis can be utilized to help prepare the assassin for his mission and to help in the planning of the mission.

A professional whose mission is to assassinate an enemy must begin by acquiring vital information about his target.

The professional should gather the following information:
   The enemy's name;
   His place of residence;
   Does he live alone;

If not, at what time is he alone;
At what time is the home uninhabited;
What security systems protect the home;
Where does the enemy work;
What is his occupation;
What security does he have during work;
Where does he go for leisure;
What security surrounds him during leisure time;
Does the enemy have martial arts training to a proficient degree;
Does the enemy possess government or military training in elite or special fields;
Does the enemy take medication;
Does the enemy use illegal drugs;
When and where can the enemy be found alone;
Does the enemy suspect a hit;
What precautions or security measures has he taken to guard against a hit.

The above information can be used to decide the time, place and method of assassination.

The professional will use the information to decide when and where the enemy will be alone, least expecting a hit and with minimum security or means of defense.

For example: If it is known that the target lives in a house or apartment alone, the professional may decide to hit him there. He must formulate a plan in order to get into the place of residence undetected. He must take into consideration any alarms or burglar detecting devices that may secure the home. The assassin must also take into consideration the possibility of neighbors becoming alarmed and endangering him and his mission. Once the assassin has entered the home he can poison any medication that the subject is taking regularly. The assassin may decide to wait for the target to return to his residence and then snuff him by means of a silent or silenced weapon. The assassin should know if the target is armed or if he has a weapon in his residence that can be a threat. In modern days bulletproof vests or body armor can be easily purchased. The assassin must know if the enemy wears body armor since it can protect the enemy and offer him a chance to defend himself.

Modern Day Ninjutsu 131

If the enemy is known to be a proficient fighter or martial arts expert the assassin would wisely choose a means of assassination that does not force him to get close to the enemy or to engage him in close combat, either armed or unarmed.

There are various methods that can be employed to snuff the enemy in his place of residence. The assassin can decide to hit his enemy by rigging explosives. The enemy's bedroom can be rigged with a timed explosive set to go off when the target is asleep. Explosives can be prepared to go off by means of a radio wave. The assassin would then wait until the enemy is in the house, call him by means of the telephone and then trigger off the explosives which were concealed near the telephone.

The enemy's bedroom can be rigged so that a lethal dose of carbintect gas will kill him in his sleep. A high power telescopic rifle can be used to assassinate the enemy. The enemy's head should be the primary target. The professional could set up a distraction which would draw and expose the enemy to a window. He would then be shot from an advantageous position.

If the assassin wanted to make the hit look accidental, he can take advantage of the enemy's use of alcohol, illegal drugs or medicine. For example, sleeping pills can be prepared so that they hold two or three times their usual amount. Once the enemy was unconscious, the assassin can start a fire in the enemy's bedroom and make it appear like it was the result of a carelessly placed cigarette. The assassin can also choose to make the fire seem like it was the result of faulty wiring such as a bare lamp cord near combustible material. In either case, the enemy would die of smoke inhalation and his death could be attributed to accident. If the enemy is caught intoxicated, he can be rendered unconscious and drowned in his bathtub. It could be made to appear that the target was intoxicated and decided to take a cold shower, slipped and fell in the tub, which resulted in him becoming unconscious and consequently drowning in his bathtub. The accidental falling of a wash cloth can be the cause for the bathtub filling up. The hit can also be made to appear as though the intoxicated enemy slipped and fell in his bathtub while preparing to bathe. The enemy would be rendered unconscious and consequently drown. If the enemy uses illegal drugs or has a radical reputation, the assassin can cause his death through overdose. The drugs can also be treated with deadly additives. It could then be made to appear that the enemy accidentally killed himself while intaking the illegal substances. If the enemy has recently suffered an emotional stress or if an emotional stress was forced upon the enemy, the assassin could assassinate him and make it look like suicide. A suicide note in his handwriting expertly forged could make the murder appear like suicide, but it is not required since many suicide vicitims do not leave notes. The enemy could be choked and left hanging from a ceiling to simulate a suicidal hanging.

An intoxicated enemy can be rendered unconscious and left in a closed garage with the car's motor running. The enemy would die of carbon monoxide inhalation and it could appear accidental. In a secluded area such as the woods or on a camping trip the enemy could be killed by food poisoning and it could be made to look like an accident. The assassin can puncture a hole in a can of food similar to that the target will be ingesting. The assassin's can would be covered until the food inside became thoroughly contaminated. The assassin would then have to mix his contaminated food into the enemy's meal. The enemy's means of transportation or communication such as the car and the telephone should be temporarily neutralized until the enemy dies. By then reactivating the means of transportation and communication and leaving, it can appear that the enemy died of botulism and did not attempt to get immediate medical help, attributing the pain to a minor illness.

The assassin must be extremely careful and knowledgeable in order to make a hit look accidental or suicidal. Modern investigative procedures and technology are very effective at determining the cause of death.

Different approaches to the hit can be followed and determined by varying target information. For example: The assassin may determine that the target is most vulnerable in the evening when he habitually patronizes a bar. The enemy can be approached and a poisoned drink offered to him in a toast. The enemy's drink can be poisoned when he goes to relieve himself or an accomplice can be used to call the enemy to the telephone or to distract the enemy while his drink is poisoned. The assassin may decide to wait for the enemy to leave the bar and kill him by means of a silent weapon or silenced weapon, as he nears his car.

Other assassination ploys are as follows:

Poison can be introduced into a target by means of powerful dart pistol or rifle similar to the ones used to catch and sedate large wild animals.

In a restaurant the target's food can be poisoned. The enemy's car can be rigged overnight to explode when the enemy attempts to start his car or the explosives can be rigged to go off after the car has been driven for a few minutes. The enemy can receive a package which contains explosives set to go off when the package is opened. The enemy can be run off the road on a dangerous cliff. The enemy can be killed by a hit and run accident. The assassin can pose as a public service employee that needs to enter the target's house to read the meter. Once inside, the enemy can be killed by a silent weapon or silenced weapon. This ploy can be useful if the enemy's house security system can not be bypassed undetected. The professional can kill the enemy and make it appear that it was for the purpose of robbing him by taking the enemy's valuables, thus concealing the true purpose of the murder.

If the enemy has bodyguards, the professional may have to terminate the bodyguards along with the target. The probabilities of body armor being used by the enemy and his guards must be seriously considered. The assassin may require assistance in the form of other well trained, trustworthy professionals. The assassin should also consider using body armor. He should know if his target has been trained and employed by the government or elite military service. A trained man such as a former C.I.A. field agent or Marine intelligence officer may have an unconscious security and may detect minor discrepancies which could result in the assassination attempt being defeated. These professionals are probably well aware of assassination procedures. The assassin should consider this possibility and plan his hit very carefully maintaining his plan as simple and effective as possible.

An equally dangerous target is one that is expecting or cautious of a hit. Be aware of the enemy's state of cautiousness and of his security preparations.

The assassin will determine the time, place and method of

the hit from the information he has gathered on his target. He must also plan the means by which to travel to and from the scene of the action. His pathways should be secured and safe. The assassin may deem it wise to prepare for unexpected obstructions or intrusions. He should be prepared to encounter more enemies at any of the three phases. The three phases being entry or meeting, the hit and the escape.

The assassin must make certain that whatever plan and means he chooses will result in the termination of the target. If he attempts a hit and fails and the target is alarmed, a second attempt could become a very difficult and dangerous task.

A field of Ninjutsu that is closely related to assassination is sentry removal. The sentry can be neutralized with a variety of weapons. A modern and effective weapon is the silenced high power rifle with a scope. Infra-red or starlight scopes should be used at night or during period of low visibility to ensure an accurate hit. The sentry can also be effectively removed with a silenced handgun. Close quarter weapons such as the knife, garrote, hand axe, entrenching tool and machete can also be employed to neutralize the sentry. The ninja must be very proficient at silently closing in with the sentry undetected in order to execute a close quarter silent kill. He must be proficient at silently walking, crawling, belly crawling, running and waiting in a concealed position unmoving. The use of camouflage clothing must be considered. The hands and face must be darkened and all loose objects secured. The ninja should consider the possibility of bypassing the sentry or guard to accomplish his mission. A missing sentry will eventually cause an alarm. The sentry must be removed when his presence is a threat to the safety of the ninja and the accomplishment of the mission.

Poisons for Assassination

The following are effective, easily acquired, easy to handle posions that can be utilized for the purpose of assassination. There are numerous other poisons that can be used to assassinate. However, they may be difficult to acquire and some require a degree of expertise to handle and employ effectively.

Botulism

Botulism is a food poisoning. It can be bred by puncturing

Modern Day Ninjutsu 135

a hole in a canned food product and allowing the contents to become contaminated. The contaminated food is then blended into the target's identical food. I.e.: contaminated beans into enemy's beans. Botulism can be treated by poison control specialists. It is best used when the enemy can not quickly reach medical attention.

Insulin

Insulin is used by diabetics to control their blood sugar level. If insulin is injected into a non-diabetic, a relatively quick death would result. The insulin can be injected through the belly button to conceal the signs of the needle. Since insulin is naturally secreted by the body the poison would go untraced unless murder is suspected and a detailed specialized autopsy is conducted.

Ethylene Glycol

Ethylene Glycol is found in some anti-freeze products. A dose of 4.5 ounces is considered lethal. The poison has a sweet taste that can be blended with sweet wines or drinks. A target dying from this poison will appear to be drunk.

Carbintect

Carbintect is found in extra strength cleaning products for spots in rugs or clothing. Placed on a pan over a heater, the resulting gas is lethal. The gas is odorless and colorless. It is best used in closed rooms with poor ventilation.

Nicotine

Nicotine can be extracted from tabacco snuff. Pour the snuff into an 8 oz. glass and fill the glass with water just enough to cover the ingredient. Wait 24 hours and filter the liquid through a handkerchief. Squeeze the snuff in the handkerchief until all the liquid has been filtered out and caught in another glass. In a pan, heat the liquid until it has evaporated into about a teaspoon of thick syrup. Nicotine will blend well with sweet wines but is best administered to a drink that the enemy is quickly gulping down. The assassin should wait if possible for the enemy to be semi or fully intoxicated before administering the poison. Half an ounce can be used for a lethal dose. It can easily kill in under twenty minutes.

Nicotine Sulfate

Nicotine Sulfate can be extracted from Black Leaf 40 and other garden insect poisons. A dropperfull in a drink will kill. The poison can also be thrown on the person. If left unwashed, the poison will be absorbed through the skin and will kill.

Other poisons which can be used to dispose of the enemy are as follows:

Tetradoxin

Tetradoxin is the poison found in the puffer or blow fish. It has been a traditional poison of ninjutsu.

Amanita Phalloides

This poison is found in the form of a mushroom. It is reputed to be one of the most lethal poisonous mushrooms.

Snake, Spider and Scorpion Venom

The venom extracted from lethal poisonous snakes, such as the Russell's Viper, the Australian tiger snake or the North American diamond back rattle snake; lethal poisonous spiders, such as the black widow or tarantula, or from lethal poisonous scorpions can be used to kill the enemy. These poisons are particularly suited for use through bio-innoculator dart pistols or rifles since it requires very little dosage to cause death.

Cyanide, Arsenic and Prussic Acid

Are very lethal poisons which can be used by the ninja. Preparation and handling of such poisons requires knowledge and proficiency and as such they should be avoided unless the ninja is an expert.

Curare and Ipoh

Curare is a traditionally lethal poison which can be used to coat darts and needles for use in bio-innoculator dart pistols, rifles or blow guns.

Ipoh is a concoction consisting of Ipoh tree juices, nicotine, garlic, snake venom, rat poisoning, scorpion tails and wasp stingers. It may sound like a witch's brew but Ipoh is extremely lethal and ideally suited for use to poison the tips and edges of weapons such as knives, swords, shurikens, tetsu-bishi (caltrops), darts, and to contaminate booby traps such as the punji stake trap.

In Modern Day Ninjutsu poisons have become a specialized and highly respected art. The more advanced practitioners recognize the validity of biological agents as poisons.

Biological agents which can be used for the purpose of assassination include STREPTOCOCCUS PYOGENES and STAPHILOCOCCUS PYOGENES. Both agents are a virulent bacteria, the latter the more dangerous, which when introduced into the body will cause septicemia or blood poisoning. Since the bacterias can be treated by medical means, another agent can be introduced to compromise the enemy by diminishing his resistance. This agent is called the Aquired Immune Deficiency Syndrome or as it is more commonly referred to by its acronym AIDS. By introducing this virus into the enemy, the ninja will effectively break down his defenses and can then bring about his death through septicemia. Depending on the dosages and virulent nature of the bacteria and by estimating the level of the enemy's diminished defense, the ninja can manipulate a rapid or slow death.

Another biological agent which can add a new dimension to poisoning is herpes genetalis. At present state medicine can not cure this herpes and as such once the vicitim is contaminated it is for life. If you couple herpes genetalis with a virulent bacteria or A.I.D.S. the ninja can manipulate a long dragged out torturous death.

Legend has it that underground freedom fighters used to contaminate rats with a venereal disease such as syphillis or gonorrhea. The rats would then consequently build up a defense but remain carriers of the disease. The freedom fighters would then release scores of the contaminated rats in the area where the enemy kept their food supply. The rats would make their way to the food and consequently contaminate it. The resulting effect was that the occupying soldiers that ate the food would contract the venereal disease. Possible applications of this and similar maneuvers introducing the newer and more serious diseases are plausible.

Biological poisoning is an advanced art which may not be

required when the mission is a rapid death by poisoning but it can be of invaluable use on special missions.

Nuclear or Radiological Poisoning is another advanced art. If the ninja is an expert in this field, he has the option of poisoning and killing the enemy through the use of radioactive thallium or other similar means. A radiological attack which can cause death through apiastic anemia, leukemia, or skin cancer can be launched with the use of portable x-ray machines. The enemy's room can be contaminated to the point where the enemy will fall sick and die. Nuclear waste dumps can provide a crafty ninja with the radioactive materials he may require for special missions, if he can get them.

Nuclear poisoning is similar to biological poisoning in that it is generally not required but it may be necessary on special missions.

Chemical poisoning has already been discussed to the degree that sodium or potassium cyanide are lethal poisons and that prussic acid is an extremely lethal poison.

Although most poisons are taken orally or are introduced via a dart or needle there is a class of poison which can be absorbed through the skin. These poisons are generally insect killers. Poisons such as: Nicotine Sulfate, Malathion, Parathion, Chlordane and Lindane can be absorbed through the skin and can cause death. The chemical dimethyl sulphoxide known as D.M.S.O. has the properties of being absorbed through the body and of carrying with it other solutions. This chemical can be used to aid in the absorption of poison through the body's pores. D.M.S.O. can be used to facilitate the absorption of the mentioned poisons through the body to result in death.

The following is a list of poisonous plants that can be used for the purpose of assassination. Some of the plants grow wild and can be acquired fairly easily. Some of the plants may be difficult to acquire. However, they are included in the list because of their lethalness and ease in handling and administering. The poisonous plants' leaves can usually be concealed and administered in a salad or sandwich.

Leaves of Rubarb — Cooked, they can kill in approximately one hour. Raw, they can kill very quickly.

Castor beans — Approximately 10 castor beans grounded up can kill relatively quickly. The beans grow wild in sections of Southern California.

Oleanders leaves are lethal.

Poinsettia leaves are lethal. They can also be used to cause temporary blindness and blisters by direct application.

Yew foliage is a very fast acting lethal substance. It is best administered in a refined state. Refined Yew foliage in gelatin capsules can be used as suicide pills.

Laurel leaves are lethal. They are best administered in a refined state.

To refine Yew foliage or laurel leaves, place the ground up foliage or leaves and 8 ounces of alcohol in a percolator. Percolate for ½ hour. Put the alcohol and what filtered through the percolator into a still. Distill off the alcohol until there is only a couple of teaspoons of residue left. Pour the residue into a saucer and let it evaporate. The remains can be used as a highly lethal poison.

Sabotage for Assassination

The following are simple basic techniques of sabotage that can be used for assassination.

How to blow up a car on ignition — To rig a car so that it blows up on ignition you need a few sticks of dynamite, an electric blasting cap, and two electric wires with alligator clips for rapid attachments. Stick the cap to a dynamite stick and connect its two wires to the electric wires. Clamp one alligator clip to the input side of ignition coil and the other alligator clip to any metal surface of the car. The explosives should be rigged near the victim.

How to blow up a car on the road — To rig a car so that it blows up on the road you need explosives such as dynamite and a fuse. Get under the hood and place the explosives as near the victim as possible and wrap the fuse around the exhaust manifold. After a few minutes of driving the exhaust manifold will get heated and ignite the fuse.

Sabotage

The field of sabotage concentrates on the destruction of property. However, it can be modified to serve the purpose of assassination. This chapter is designed to present the basics on the art of sabotage. The following are sabotage weapons.

Gasoline and Fire
Gasoline is an excellent weapon for sabotage. The intended target is drenched with gasoline and ignited with a match, lighter or torch. If a delayed reaction is desired, a fuse can be rigged with a timer. For example, a gasoline soaked thin rag can be placed under a cigarette. A cigarette usually burns an inch per seven minutes. If the fuse were located an inch in from the end, it would take approximately seven minutes before ignition.

Modern Day Ninjutsu

The gasoline and fire can be used as a means of assassination by burning a room where a trapped or unconscious enemy lies or by soaking the enemy with gasoline and setting him on fire. An ignited gas soaked rag stuffed into a car's gas tank can be a means of sabotaging an automobile.

Fire Bombs

A fire bomb consists of a bottle filled two thirds with gasoline. The mouth of the bottle is fitted with a gasoline soaked rag. To use, ignite the rag and throw the bottle. When the bottle hits the pavement or wall, it will break and the gasoline will ignite. The fire bomb is best thrown against hard surfaces. A mixture of one part oil, two parts gasoline can be used instead of just gasoline.

Dynamite

Dynamite can be of use in blowing up cars, buildings, bridges, etc. Dynamite sticks can be detonated by means of a blasting cap and a fuse or by electronic detonators.

There are numerous other chemicals and explosive devices which can be used for sabotage by the ninja with expertise in explosives and demolition. They include either commercial or homemade: purified ammonium nitrate, gunpowder, plastique explosives, nitroglycerine, and trinitrotoluene (T.N.T.). Also homemade blasting caps can be constructed using mercury fulminate. **NOTE:** *these explosives require expertise in their construction, care and usage.*

Espionage

The field of espionage concentrates on information gathering, interpretation of the information, then action based on the interpretation. The art of espionage can be in effect at a very sophisticated level such as in our Government agencies and Military forces, or it can simply involve one man. Espionage can encompass a diversified range of missions. Sabotage and Assassination can be missions that are carried out to fulfill the demands of intelligence. Sabotage and Assassination both also require the use of intelligence procedures to acquire the information necessary to carry out the mission.

Knowledge about the enemy's previous activities can be used to lead the agent in acquiring and interpreting present information. Through the use of logic and reasoning, the information can be analyzed and interpreted. Creative thinking checked by logic, can be used then to help formulate an adequate plan of action.

The basic guidelines that an agent must understand and follow in order to carry out espionage missions such as observations, reconnaissance, document acquisition or duplication, assassination or sabotage are as follows:

Know and understand mission

The agent must understand the nature and purpose of the mission. He must be fully aware and positive of his duties and responsibilities for the mission.

Chain of Command (if applicable)

The agent should know his team members and acknowledge a team leader. The agent should know who he is to report to and receive orders from as well as who reports to him and receives orders from him. The above does not apply if the agent operates alone.

Communications (if applicable)

The agent should have means of communication with his team members and if necessary, with his superiors. Portable two-way radios and signal mirrors can be used. There should be a primary and alternate means for communication.

Mobility

The agent should have the means to transport himself, team members and supplies to and from the site of his mission. The means of mobility can include a prepared automobile or walking. If the agent must walk a great distance, he should prepare himself physically as well as acquire appropriate footwear, supplies and backpacks.

Cover

The agent should acquire an excellent cover so that he can operate undetected. This can range from taking an office job position to proper camouflage cover and concealment in the field.

Supplies and Equipment

The agent should acquire and prepare the necessary supplies and equipment that will be needed for the mission.

Training

The agent must prepare for his mission by enduring the appropriate training. If a specialized skill is necessary, then a trained expert team member must be recruited for the mission.

Opportunity

The ninja must plan his mission so that he takes advantage of opportunities presented that will offer and enhance his chances of success.

Espionage missions can be of general or specific nature. For example, an agent can be sent to observe the general activities of a camp or office without knowing exactly what he is looking for, or the agent can be sent to observe a specific individual to acquire information that will prove he is an enemy informer.

General Field Observation

An agent's mission can be to do some field observation on an enemy force. The agent should be trained to observe the specifics which may be of interest and which may be of value. The following are some points which should be recorded for evaluation: terrain, organization of enemy forces, strength and number of enemy forces, identity of unit, movement of enemy, training back-up or support forces, supplies, supply system and procedures, chain of command. The agent should also be proficient at sketching and constructing a valid map for analysis and future use.

The observations can be carried out from a fixed advantageous position or the agent may be forced to track the enemy's movements.

Specific Missions

Specific missions can require the skill of following a person undetected, of setting a stake-out or of covert entry, search seizure and duplication.

Surveillance

If an agent must follow a subject undetected, he should make great efforts to blend in with his surroundings. Team members can be used to facilitate the "tail." Inconspicuous movements, such as, lighting a cigarette or looking at the wristwatch can be used as a signal to communicate with the team members. It is a general rule in surveillance that it is better to lose the subject than to have him become alarmed and aware of the tail. The subject can be followed closely or at a greater distance. He should never be contacted in any manner while he is being tailed.

Auto surveillance

The ninja may be forced to tail a subject by automobile. Various automobiles should be used to facilitate the tail. The team members would follow on side streets and pick up the subject if he turns. Two-way radios can be used for communication.

Stake outs

An agent may be forced to position himself in an advantageous position to observe the enemy. The ninja should make great efforts to remain anonymous and inconspicuous. He should have an excellent cover to remain undetected.

Covert entry, search, and seizure or duplication

The ninja's mission can require the covert entry into an office, room or house. He should have the means by which to enter the room. If locks have to be picked, the agent should be proficient at lock picking or have an expert team member that is so qualified. A detailed search may be called for. The ninja should be trained to search the area and to look for hiding places without disrupting the area. Once found, the material or documents can be seized and removed or duplicated by means of a camera or if necessary, short-hand and sketches.

In the planning and execution of his mission, ninjas should be aware and alert for tails, security and electronic surveillance. Plans should be drawn up to deal with the above dangers.

Specific training

The ninja may have to endure specific training for his missions. The training can and should incorporate the language, culture and traditions of the enemy or population that will be contacted or engaged. Other specific training can include field crafts, reconnaissance procedures, lock picking, and neutralizing electronic surveillance equipment.

Tactical considerations

The ninja may deem it advantageous to befriend an unsuspecting friend of the targeted enemy in order to accomplish his mission.

If the ninja is discovered or placed as an informer, he can release false information to mislead the enemy. He should first gain the enemy's trust by releasing declassified or unimportant true information.

Escape and Evasion

This chapter will cover some technical and strategical considerations for escape and evasion. A ninja may be forced to evade pursuit or to escape capture in order to accomplish his mission. The professional will plan his escape route and prepare for enemy interference. The exit route can be the same as the entrance or it can be a different one. The ninja can enter and leave through the same window or door or he may consider it more practical to enter through one door and leave through a rear door or window. The ninja may find it necessary to rapel into or out of a building's window. The exit route should be thoroughly known to the ninja. He should know where he can quickly dodge behind cover to confuse and mislead the enemy. He should know where he can stop behind adequate cover to fire upon an exposed and unprotected enemy. The ninja can use the traditional poisoned metal caltrops to stop his pursuer. The caltrops are best used on an enemy giving chase in an area where the caltrops will remain undetected until it is too late to avoid or defend against them. He can lead his pursuers into an ambush.

In his escape plan the professional can consider boobytraps or mining his escape route so that enemy pursuers can be trapped and stopped. Make certain that the traps do not disturb the surroundings and that they can pass undetected until they are in effect. Even if all the set traps are not triggered, an effective or discovered trap will serve as a psychological weapon. The enemy can become demoralized and extra cautious, allowing the ninja the time or delay needed for his escape. Traps or mines should be set in an area where the pursuers will surely traverse and where the triggering of the traps or mines will offer the best chance for an escape.

Some traps that can be used to stop pursuers are the following:

Punji Stake Trap

The punji stake trap is constructed by digging a 2x2 meter hole 2 meters deep into the ground. Boards or logs are then placed in to hold the dirt from refilling the hole. Punji stakes or thin metal slivers are positioned on the bottom to impale the enemy. The stakes or slivers can be poisoned or contaminated with blood and dung to kill or severely demotivate the enemy. The hole must be covered and camouflaged. A thin sheet of plastic secured to the walls can be used as a roof. The original dirt, debris or grass can then be placed on the sheet to conceal and camouflage the trap. The punji stake trap is best positioned in an area where the enemy has poor visibility such as behind a fallen log or in an area where the enemy will be moving in a hasty, uncautious manner.

Metal Spike Trap

The metal spike trap is constructed by digging a hole 30x30 centimeters wide, 100 centimeters deep. Boards can be placed to prevent the hole from refilling. Sturdy metal slivers or spikes sharpened to a fine point are then attached to the boards six centimeters from the top. The spikes are positioned in 40 to 45 degree angles. The trap must then be concealed. When the enemy steps on the camouflaged roof his foot and leg will fall in. The spikes will enbed themselves into his leg. If the enemy tries to withdraw his leg, the spikes will dig in deeper. The spike tips can be poisoned or contaminated.

Explosive Light Bulb

A light bulb can be drilled and injected with gasoline to produce an explosion. Rig the prepared light bulbs in rooms or passageways where the enemy is likely to turn on the lights if he suspected an intrusion. The bulbs must be set in places where the enemy will receive the explosive effects while the ninja remains in a safe area.

Claymore Mines Trap

If the professional is proficient with claymore mines and he can obtain them, the mines can be set in an area where the enemy will pass when in pursuit. The mines can be detonated by the concealed agent.

Handgrenade Traps

If the ninja is proficient with handgrenades and he can obtain them, they can be used to slow down or stop a pursuing enemy. The grenades can be thrown in the conventional manner or they can be set as traps by pulling the pin and placing the grenade in a secure can that will maintain the safety lever pressed, therefore not activating the fuse. A strong thin wire can be tied to the grenade and the other end tied to a tree or doorknob. When the enemy unknowingly forces the wire by walking through it or opening the door, the grenade will be pulled out of the can and the fuse will become activated. The professional must be aware of the time it will take to detonate his choice of grenades. Cacodyal and makeshift fire grenades can be similarly employed.

Ambush Trap

The ninja can lead the enemy into an ambush. The professonal's teammates can be in a concealed area and open fire on vulnerable, unsuspecting pursuers. The ambushers should designate a kill zone that will allow the ninja to get to safety while leaving the pursuers out in the open with little or no cover. The ambushers must be proficient with the semi or fully automatic weapons of their choice. Claymores and grenades can be used to supplement the firepower.

Fire Trap

The ninja can have some incendiary material set near a door or passage concealed in a doormat or garbage can. He can ignite the material with an incendiary grenade or by other means such as lighting a fuse. The resulting rapid fire can delay or prevent the enemy from pursuing.

Evasive Driving

Evasive driving is a skill that can be of value to the modern ninja. The ninja's escape plan can eventually entail driving away in a car. The professional should consider preparing his automobile and his driving skills to meet the potential demands of evasive driving. The first consideration is the choice and preparation of the automobile. The car should ideally be powerful and have excellent handling. It should be capable of hitting high speeds while maintaining excellent handling. The Porsche 914 or 911 and the Datsun 280Z Turbo are examples of automobiles that can move quickly while maintaining stability. These cars can cut in and out of traffic and execute sharp turns with relative ease. These cars, however, are not ideally suited for ramming. If the ninja feels that he requires a car with a powerful engine and a body that can be used for ramming then an older American-made car can be chosen. The car should have a metal frame and a large engine. These cars will be fast on a straight road and can be used to ram through roadblocks. Whatever the choice, the car must be in excellent mechanical condition and have excellent tires. The installment of stabilizing bars can be considered to help stabilize the car. The ninja can choose to have his car bullet proofed, but the effects of excess weight in the car's handling should be noted, however.

If the evader has a firearm or an accomplice with a firearm, shots can be directed at the enemy car's front tire on the driver's side.

One important point to remember when in an evasive driving situation is to avoid crashing at all costs. If the operator

Modern Day Ninjutsu 151

crashes, even if he is not seriously hurt, he will most surely be captured. The concept of evasive driving is to make moves and turns that the pursuer can not follow. For example, if the ninja is travelling north and pursued from the rear, in a busy intersection he can execute a sharp left turn cutting barely in front of the oncoming traffic. The pursuer may be forced to stop to avoid getting hit. The maneuver can offer the evader a chance to escape.

The following are evasive driving techniques that can be used to avoid capture. The maneuvers should be practiced with the chosen car to acquire expertise and to determine if the car is suited for the maneuvers.

The Moving 180

While moving at speeds of approximately 30 mph, release the gas pedal, turn the wheel so that it is ½ of a full turn while simultaneously hitting the emergency brakes. If you drive a standard then you must press on the clutch. You should not be past second gear; if you are, then you must take the car out of gear. When your car has turned approximately 90 degrees, release the emergency brakes, step on the accelerator and steer. If the car is not in gear you must also throw it in first or second gear.

The Reverse 180

While moving in reverse at speeds of 25 to 30 mph, release the accelerator while quickly turning the steering wheel to the left. If you are driving a standard shift you must press on the clutch. When the automobile is at an approximate 90 degrees, shift into first or second gear and drive away.

Ramming

It is preferable in an evasive situation to avoid and outmaneuver the pursuing or ambushing car. Ramming should only be used when there are no alternate means of escape. The evading driver should know the sturdiness and power of his car for this maneuver. Wear the seatbelts at all times. Ramming should be executed at speeds of approximately 20 to 30 mph. Try to hit the target car at an angle, keeping the accelerator depressed during impact. The most favorable area to ram through is the target car's rear bumper and wheel. The front bumper and wheel area can also be rammed.

If the purpose of ramming is to run the enemy's car off the road it can be accomplished as follows. From the rear, with your front right bumper, ram the enemy car's rear left bumper at an angle. This maneuver can force the enemy's car to skid and veer to the right. If you are side by side and slightly ahead of the enemy car, you can run him off the road by hitting his car's left front bumper with your right side.

Smoke Screen

The evasive driver's car can be prepared so that it will produce a smoke screen to blind and confuse the pursuers. To prepare the car, drill a hole the size of a paint sprayer's nozzle in the exhaust manifold. Weld the nozzle in place and secure a length of gas line to the nozzle tube extending the other end of the line into the driver's compartment. Attach a spray unit to the extended line. The spray unit can be filled with castor oil, preferably burned. To create the smoke screen squirt the oil so that it runs down and hits the exhaust pipe.

Escaping Capture

If the ninja is captured and arrested he may still have a chance for escape. He must remain very alert while seemingly subdued to take advantage of any escape opportunities. The ninja should be prepared to loosen himself from rope ties or handcuffs in order to escape. A concealed razor blade secured by means of a hidden pouch or sewn to the inside of the trousers waistband near the center of the back can be used to escape. If the hands are tied behind his back he can easily reach in and withdraw the razor. If the hands are secured in front of him and he can not reach the tool, he may have to unbuckle and lower his trousers in order to withdraw his tool. The ninja can choose to conceal numerous razor blades on his person for use in escapes or as makeshift weapons. He should wait until he has an opportunity and with the razor blade cut through his bonds. Concealed razor blades can be used to escape from all standard rope ties including the effective hogtie.

The ninja should also be prepared to escape from handcuffs by carrying a set of picks or master key. The picks or keys can be concealed in the inside of the pants waistline in a concealed pocket for easy access when under restraint. The tools can also be concealed in boots. Some handcuffs can be shimmied by using a thin piece of wire. The wire is inserted along the end edge of the shackle and the catching mechanism is forced backward. This will produce a smooth surface and the catching teeth will not lock allowing the shackle to be pulled free. Handcuffs can be opened with a bobby pin used as a modified pick. If a torsion wrench is required for picking, a pen's pocket clip bent outward at approximately 90 degrees can be used as an improvised torsion wrench. Many similar handcuffs can be opened by the same master key. If the ninja knows which types of handcuffs the enemy uses he can acquire a master key. Pick tools or master keys can be concealed in a modified wrist watch.

Ninja climbing walls using rope ladder and hooked Bo or staff.

Shimming open standard handcuffs behind your back is a valuable escape maneuver to the Modern Day Ninja.

A running eye-gouge takedown can be used to quickly drop an enemy while on the run.

Conclusion

Ninjutsu covers a broad spectrum of study. It is a disciplinary and military art which is constantly adopting the possibilities which may exist for future employment while maintaining the underlying qualities and factors which have allowed it to become one of the most proficient, feared and respected combat and survival arts. Ninjutsu has been progressively advancing through the ages. For example: the traditional assassination ploy by which the ninja would conceal himself in the warlord's latrine and then impale the warlord through the rectum with a spear when he went to relieve himself is compatible with the modern-day ploy of rigging a chair so that when the victim sits on it, it collapses protruding a hollow glass knife impaling the victim through the rectum; or with the ploy of rigging the urinal grill so that the victim electrocutes himself while urinating. Similarly the ninja's traditional weapon, Shinobi Zue or hollowed staff which would conceal a spear head or chain has been replaced by walking canes which in reality are concealed shotguns. Many more similarities can be drawn between the special weapons and tactics of traditional ninjutsu and those found in the former O.S.S., the C.I.A. and special commando units.

Although weapons, tactics, procedures and ploys may change in modern and future days, the underlying concepts of stealth, invisibility, ruthlessness and practicality remain the same.

Appendix
— A Word on Your Personal Weapon —

The ninja should choose and acquire a personal weapon. The choice can range from an M-16 automatic rifle to a ball point pen. The choice will be determined by the nature and purpose of the weapon and the restrictions in effect. Once the ninja has chosen his personal weapon, he should adhere to the following rules.

    1 — **Keep it with you at all times**

    A personal weapon should be on your person at all times with little or no exceptions. It should be considered a part of your body or clothing. A weapon that is left at home, base or car is useless if you must defend yourself unexpectedly.

2 — Master the weapon

The ninja should become thoroughly familiar with his weapon and its uses. He should know how to instantly disable an enemy with his weapon and how to use it in hand-to-hand combat.

3 — Become lightning quick

The ninja should practice and become very quick with his weapon. A lightning-like attack can offer him a chance for survival against great odds.

4 — Use it reflexively

The ninja should train himself to reflexively reach for his weapon when in danger. By being aware of the weapon and its uses, he will eventually consider the weapon as an extension of his person.

5 — Learn to use it against other weapons

The ninja should learn how to combat against other weapons utilizing his personal weapon. In some cases, he may have to rely on attacking first and quickly disabling the enemy. In others, he may be forced to engage the enemy in close combat. He should know how to neutralize the enemy's weapon(s) or advantages while executing his attack in order to survive.

A personal weapon can be the difference between life or death. A quick thrust of a pen through the enemy's eyeball can neutralize even an enemy armed with a handgun. A small folding pocket knife can be used to combat multiple adversaries. A personal weapon and the skill needed to wield it effectively is an important lesson taught in Modern Day Ninjutsu.

OTHER AVAILABLE TITLES

☐ **HOW TO MAKE DISPOSABLE SILENCERS.** Detailed manual includes over 65 close-up photos and drawings to help anyone build disposable silencers, that can be constructed in seconds, without tools, using only inexpensive, readily available items.
No. 25 $12.00

☐ **FULL-AUTO!** Newly updated edition! Complete illustrated manual on selective fire conversions for the following weapons. MINI-14, AR-15, HK 91-93, MAC 10-11, and the M1 CARBINE.
No. 21 $10.00

☐ **IMPROVISED MUNITIONS HANDBOOK.** The most sought after Army manual ever published! Includes revealing chapters on: Improvised explosives and propellants, mines and grenades, small arms and ammo, mortars and rockets, plus much more!
No. 30 $12.00

☐ **TWO COMPONENT HIGH-EXPLOSIVE MIXTURES AND IMPROVISED SHAPED CHARGES.** Complete details on the production of explosives easily made by combining two compounds which—when separate—are completely safe, yet when mixed produce some of the most powerful non-nuclear explosives known. Also provided is an in-depth look at improvised shaped charges.
No. 78 $8.00

☐ **PYROTECHNICS.** With over 245 pages, this monumental work places in the hands of the beginner a working manual which will assist greatly in the production of every known piece of fireworks.
No. 45 $10.00

☐ **HOW TO MAKE DISPOSABLE SILENCERS, VOL II.** More sophisticated silencer designs, which can be constructed with a minimum amount of time and money, without the need for expensive tools. Silence weapons of any caliber (from .22 LR to 30,06). Step-by-step directions, as well as over 100 close-up photos are included to make the job easy.
No. 74 $14.00

☐ **9mm SUBMACHINE GUN.** Complete, full-scale drawings and written instructions to allow the average individual to construct a 9mm submachine gun without a machine shop.
No. 20 $5.00

☐ **IMPROVISED WEAPONS OF THE AMERICAN UNDERGROUND.** Completely explains how to make nitroglycerin, plastic explosives, detonators, silencers and many others. Also included: complete plans for a homemade submachine gun.
No. 42 $7.00

☐ **LOCKS, PICKS AND CLICKS.** The manual currently being used to train CIA personnel in surreptitious entries and searches. It details various search techniques, simple methods of entrances and openings without keys, lock picking, and safe manipulations.
No. 54 $6.50

☐ **THE LOCK PICK DESIGN MANUAL.** The most detailed manual on lock pick design available to the public. This book has been used by various Government agencies as both a textbook and to assist in surreptitious entry.
No. 55 $6.00

Please send me the book(s) I have checked above. I am enclosing a check or money order for $ _____. Sorry, no COD's.
☐ Send FREE catalog.

NAME _____

ADDRESS _____

CITY _____ ST. _____ ZIP _____

Mail to:
    J. FLORES P. O. Box 14, Dept. N , Rosemead, CA 91770